BIOLOGICAL AND MEDICAL PHYSICS
BIOMEDICAL ENGINEERING

BIOLOGICAL AND MEDICAL PHYSICS
BIOMEDICAL ENGINEERING

The fields of biological and medical physics and biomedical engineering are broad, multidisciplinary and dynamic. They lie at the crossroads of frontier research in physics, biology, chemistry, and medicine. The Biological & Medical Physics/ Biomedical Engineering Series is intended to be comprehensive, covering a broad range of topics important to the study of the physical, chemical and biological sciences. Its goal is to provide scientists and engineers with textbooks, monographs, and reference works to address the growing need for information.

Continued after Index

Matthew Simon

Emergent Computation
Emphasizing Bioinformatics

With 274 Illustrations

Matthew Simon
Notre Dame de Namur University (retired)
Belmont, CA
USA
plasmid@cybcon.com

Library of Congress Cataloging-in-Publication Data
Simon, Matthew
    Emergent computation : emphasizing bioinformatics / Matthew Simon.
      p. cm. — (Biological and medical physics, biomedical engineering)
    Includes bibliographical references and index.
    ISBN 0-387-22046-1 (hbk. : alk. paper)
      1. Biomedical engineering.   2. Bioinformatics.   3. Medical informatics.   I. Title.
    II. Series.
    R856.S47 2004
    572.8—dc22   2004049168

ISBN 0-387-22046-1   Printed on acid-free paper.

Printed in the United States of America.   (SBA)

9 8 7 6 5 4 3 2 1   SPIN 10994054

springeronline.com

Series Preface

The fields of biological and medical physics and biomedical engineering are broad, multidisciplinary and dyanmic. They lie at the crossroads of frontier research in physics, biology, chemistry, and medicine. The Biological & Medical Physics/Biomedical Engineering Series is intended to be comprehensive, covering a broad range of topics important to the study of the physical, chemical and biological sciences. Its goal is to provide scientists and engineers with textbooks, monographs, and reference works to address the growing need for information.

Books in the series emphasize established and emergent areas of science including molecular, membrane, and mathematical biophysics; photosynthetic energy harvesting and conversion; information processing; physical principles of genetics; sensory communications; automata networks, neural networks, and cellular automata. Equally important will be coverage of applied aspects of biological and medical physics and biomedical engineering such as molecular electronic components and devices, biosensors, medicine, imaging, physical principles of renewable energy production, advanced prostheses, and environmental control and engineering.

Elias Greenbaum
Oak Ridge, TN

Preface

While teaching courses in automata theory, I was a little dismayed to find that students were under a misapprehension, a prejudice concerning the subject. My students thought that automata theory was applicable only to the studies of computer science, engineering, and mathematics. I sought interesting examples of automata theory in these traditional areas, as well as in non-traditional areas to dispel this view.

The purpose of this book is the examination of non-traditional applications of automata theory not normally encountered when studying computer science, engineering, and mathematics.

Varied, unusual applications of automata theory could be found in areas of emergent computing, such as bioinformatics, as well as in sociology, anthropology, biology, chemistry, geology, philosophy, psychology, medicine, etc. Indeed, automata theory is ubiquitous.

As I carried out research in the areas of emergent computing, another viewpoint emerged. Specifically, the view that most emergent applications of mathematical linguistics did not deal with human communications. The prejudice that language is limited to human beings has not escaped the attention of others.

Writing this book presented difficulties. The first problem was that no book has attempted to be so comprehensive. There was no model for this book. Thus I had to effectively create a new area of study. Second, as the subject matter involves multiple disciplines, the book had to be comprehensible to readers from different educational backgrounds, yet not be unreasonably demanding. The chief disciplines include organic chemistry, biochemistry, genetics, biology, and theoretical computer science. Thus, I felt that a brief review of some aspects of organic chemistry and biochemistry should be included sufficient to support comprehension of these disciplines. A brief review of automata theory is also included as it is hoped that biochemists would be interested in automata theory. Third, as emergent computation is rapidly expanding, keeping up with the vast and rapidly growing research data is difficult.

It is hoped that this book will be found useful and interesting to those studying emerging, non-traditional applications of automata theory. Each chapter will have an interesting historical note relating to medicine or some related subject. It is hoped that the university environment will become less adverse to interdisciplinary studies: censorship is not compatible with discovery and research!

Last, I take full responsibility for errors and especially omissions.[a] In many cases, I have discussed only some of the topics covered in research papers. It is expected that interested readers will read further to gain greater insights.

I dedicate this book to my mother, Tomar Rosenberg, my wife, Esther M. Lederberg, and Randy, Sarah, and Nicole.

Matthew Simon

---

[a] I am especially apologetic that I could not find applications of mathematical linguistics applied to cetaceans, primates other than human beings, social insects other than bees, geology, psychology, etc. I sought such papers, contacting specialists, but I was not able to locate appropriate papers.

Introduction

This book has as its object the study of emergent computation. Bioinformatics is emphasized, but the subject matter is not limited to bioinformatics.

Insofar as bioinformatics is examined, DNA, RNA, and proteins are studied. While this might appear to be natural and expected, the reader should be aware that other studies in bioinformatics exclude some or all of these subjects!

It might also come as a surprise to the reader that other studies in bioinformatics emphasize computation theory to such a degree that they ignore biochemistry! These are works of computability that unfortunately ignore and are not compatible with the known facts about DNA, RNA, and proteins. Apparently the view held by these authors is that their theories of computability are so beautiful that DNA, RNA, and proteins are less important than elegant theories. This book will endeavor to study the bioinformatics of DNA, RNA, and proteins based upon published papers of research scientists. Speculative theories of computation occasionally applied to limited aspects of bioinformatics are omitted.

There are difficulties presented in attempting to cover so many aspects of bioinformatics. Bioinformatics is now a very large study, and demands much of both author as well as readers. Readers must be knowledgeable in a number of areas:

- Inorganic chemistry, biochemistry, organic chemistry [b]
- Automata theory
- Aspects of biology (such as microbiology, and genetics)

It is possible to study only portions of the subject matter covered in this book, depending upon time and interests. A knowledge of chemistry is not essential, but there will be a loss of appreciation and comprehension of bioinformatics without such knowledge.

An interesting fact emerges when one studies the subject of emergent computation, a fact that should not be ignored: bioinformatics is not entirely new–this subject has a history.

The geologist James Hutton (1726-1797) noted that geological structures are subject to change. These changes are due to high pressures and extreme temperatures on land and under the sea, erosion, and volcanic action: rock on dry land becoming submerged, then re-emerging in a cyclical process, over long periods of time. This theme was picked up by Charles Lyell (1797-1875), and modified by Uniformitarian views that the same scientific laws acted uniformly in the past as they do in the present. The gradual geologic changes seen by Hutton, supported by a paleontological record (fossils and sedimentary strata), with an appreciation of the immense age of the earth, came in conflict with the prevailing religious views of Catastrophism. According to Catastrophism the planet Earth was created in a short span of time, the earth and all its creatures created in an instant, by an

---

[b] The author has excluded as much chemistry as possible, such as qualitative analysis (inorganic and organic), quantitative analysis (titrimetric and gravimetric, and instrumentation, such as chromatography, gel electrophoresis), thermodynamics, physical chemistry, and most of genetics. Cellular automata, neural nets, and Post systems are not covered. Some studies are limited to theory, while others extend to in vitro or in vivo implementations. Bioinformatics does not focus upon database mining, nor the alignment of sequences of nucleotide bases in this book.

omnipotent and omniscient being (or in a modern guise: altered by the external effects of asteroids[c]). The planet thus created then remains fixed and unchanging. Darwin and Wallace (*On the Origin of Species by Means of Natural Selection, or the Preservation of Favoured Races in the Struggle for Life*, 1859) extended the conception of the evolution of inorganic matter, understood by these pioneering geologists, to the evolution of life forms by natural selection. Thus evolution moved from being applied to inorganic matter into the realm of organic living organisms. This was a revolution in thought, a revolution still opposed by many people who continue to object to scientific inquiry.

This brief view of the history of evolution is interesting because of the progressive application of evolution from inorganic matter to organic matter. This history of evolution has been forgotten, and is now unknown to many people. Many people think of evolution as only applying to organic living things. That evolution applies to inorganic matter seems strange, even novel.

An interesting parallel exists in bioinformatics. The common prejudice is that language is an attribute limited to living things, specifically human beings.[d] Lindenmeyer systems are the study of language or automata theory applied to describe plants, bacteria, sea shells, etc. Bioinformatics also extends the study of language from human beings to non-human subjects such as biomolecules. Thus the process is reversed, language being extended from human beings to biomolecules, in opposition to the common prejudice. Indeed, many people who study language deny that language theory is a property of molecules: rather it may be "language-like" but certainly it is not language [165]! If one defines language as requiring a "conceptual-intentional" system, as well as "sensory-motor systems" (auditory processing and speech processing), then indeed, language may be limited to humans [31]. However, this anthropocentric view of language is rather biased. A view that ignores the results of scientific investigation to maintain anthropomorphic prejudices in the name of science is nothing more than a new religion. Language may also be viewed as merely a way to communicate observed patterns (patterns of molecular forces). "Communication" need not imply a psychological consciousness. Indeed, when we refer to communicable diseases we recognize this fact. The complementary nature of nucleotide bases can be "communicated" through natural agencies other than conscious thought, specifically through molecular forces. The view that language requires auditory and speech structures, as well as a conscious psychological state is only one view. Similarly it has been proposed that birds use language to communicate sexual selection. Once again, communication need not imply a conceptual-intentional system. Many authors of research papers referred to in this book have an alternate view of language in which DNA, RNA, proteins, bird calls, etc. may be accurately classified as languages.

It is my hope that readers will enjoy this book in spite of errors and omissions.

---

[c] This includes floods of lava (Plutonism) or water (Neptunism), mountain range formation, earthquakes, or modern non-linear chaos theory [111, pp. 36, 37].

[d] The idea is that language is limited to human beings, that language is in fact the essence of man [111, p. 45].

Table of Contents

Part I

Emergent Computation: Bioinformatics

The objective of Part I of this book is to study different aspects of the interrelationship between automata theory, also known as the theory of computability or mathematical linguistics, as applied to major areas of bioinformatics: DNA, RNA, and protein chemistry.

Chapter 1:   A Review of Chemistry

Medicine

"Despite the presence of faculties of medicine in most universities, major centers of academic medical education were always few in number. In terms of chronological and geographical distribution of such centers, the salient facts are the early emergence and early decline of Salerno (the late eleventh through the early thirteenth centuries) and the subsequent predominance of just three centers: Bologna, Montpellier, and Paris. A fourth center, Padua, the home of medical studies since the thirteenth century, increased in size and importance in the course of the fifteenth century.

"The differences between the various faculties of medicine lay more in size, reputation, and institutional position than in curriculum. The common heritage of Greek and Islamic medical learning and Aristotelian natural philosophy ensured a good deal of curricular uniformity throughout Europe, but the intellectual as well as the professional and collegial ambiance of the various medical faculties varied widely.

"The introduction of an academic and book-oriented emphasis in Salernitan medicine was the result of a confluence of factors. Access to some of the medical writings in Latin available in Italy during the Middle Ages no doubt played a part. In addition, the geographical situation and early medieval history of the southern part of the Italian peninsula provided opportunities for intellectual contact with both the Greek and Arabic speaking worlds. Extremely important was the presence nearby of the celebrated Benedictine monastery of Monte Cassino, where Constantinus Africanus was a monk. As noted...Constantinus' *Pantegni* and other translations provided the first access in western Europe to a substantial corpus of medical literature derived from Arabic sources.

"In addition to writing various guides to medical practice, twelfth and early thirteenth-century Salernitan authors brought together a collection (subsequently known as *articella*) of short treatises conveying the rudiments of Hippocratic and Galenic medicine to serve as a basic curriculum, and they established the practice of teaching by commentary on these texts. The collection as first compiled in the twelfth century (later other texts were added) consisted of two Hippocratic treatises, the *Aphorisms* and the *Prognostics*; a brief Galenic treatise known under various titles (*Ars medica, Ars parva, Tegni,* or *Microtechne*); an Arabic introduction to Galenic medicine known to the Latins as the *Isogoge* of Johannitius; and short tracts on the main diagnostic tools of the medieval physician, namely pulse and urine. The association of medicine with natural philosophy was also emphasized at Salerno; Salernitan masters were among the earliest Latin writers to reflect some knowledge of Aristotle's writings on physical science, and the well-known 'Salernitan questions' mingled medical and general scientific topics"[231, pp. 55-58].

"In addition to, and often as a part of, the rise of the boards of health came another phenomenon of the post-plague era–the plague doctor. In more sophisticated parts of the West, beginning in Italy and then spreading to France, England, the Netherlands, and Germany, town councils or health boards hired municipal physicians and surgeons to treat plague victims. It was a difficult, dangerous, and unpleasant job, and, to make matters worse, after the plague doctor had treated the victims, he had to endure a long quarantine of his own. Who, then, were these doctors, and why would anyone want the job? Very few were established doctors. Usually, the job was filled by second-raters who had difficulty establishing practices of their own, or by young physicians and surgeons, generally from rural areas, who were just starting out" [227, p. 125].

Organic Chemistry and Biochemistry

This chapter constitutes a brief discussion of some of the basic ideas and nomenclature of biochemistry and aspects of biology. The knowledge gained from studying this chapter is required for subsequent chapters. This chapter is not intended to impart extensive knowledge of organic chemistry, biochemistry, or biology, but merely the most elementary basics. For those readers with little background in these subjects, this chapter should serve as an aid to comprehension. If more extensive knowledge is desired, please refer to referenced books or articles.

A Brief Review of Chemistry

Bonds, Lewis Acids

It is not the purpose of this book to delve deeply into chemistry; thus only brief mention will be made of topics such as bonds. The interested reader is welcome to a deeper study in books covering chemistry. It is sufficient for the purposes of this book to mention that bonds of interest include polar (ionic) bonds and covalent bonds, understanding that these two kinds of bonds represent extremes. Bonds may take on the nature of something between both these extremes, a "percentage" but even saying this goes beyond the purpose of this book. An atom that can fairly easily provide (a mobile) electron (thus is negatively charged) to an atom deficient in an electron, will participate in an ionic, or polar bond. When two electrons are shared or resonate between two (or more) atoms, these electrons partake in a covalent bond. We shall not delve into the molecular orbitals that these electrons occupy. In addition to polar and covalent bonds, there are dative bonds. Acid or salt structures may be viewed as Lewis acids, but once again, this extends somewhat beyond the purpose of this book.

Chelates

Chelates [8] will receive attention; one specific area of interest will be found in the chapter dealing with nucleotide bases in DNA, discussed in Chapter 3.

reaction rate

$$M + L \longrightarrow ML \qquad k = \frac{[ML]}{[M][L]}$$

M refers to a cation (often a metal)
L refers to a ligand anion
[ ] refers to a concentration

What then is the reation rate for : $M + 4L \longrightarrow ML_4$?

Stepwise Formation Constants

$$M + L \rightarrow ML \quad , \quad k_1 = \frac{[ML]}{[M][L]}$$

$$ML + L \rightarrow ML_2 \quad , \quad k_2 = \frac{[ML_2]}{[ML][L]}$$

$$ML_2 + L \rightarrow ML_3 \quad , \quad k_3 = \frac{[ML_3]}{[ML_2][L]}$$

$$ML_3 + L \rightarrow ML_4 \quad , \quad k_4 = \frac{[ML_4]}{[ML_3][L]}$$

then:

$$[ML] = k_1[M][L]$$

$$[ML_2] = k_2[ML][L] = k_2 k_1 [M][L]^2$$

$$[ML_3] = k_3[ML_2][L] = k_3 k_2 k_1 [M][L]^3$$

$$[ML_4] = k_4[ML_3][L] = k_4 k_3 k_2 k_1 [M][L]^4$$

Thus we conclude that

$$k = \frac{[ML_4]}{[M][L]^4} = k_4 k_3 k_2 k_1$$

Equilibrium Constant

Also, if we consider the following reaction:

$$A \;+\; B \;\underset{k_2}{\overset{k_1}{\rightleftarrows}}\; C \;+\; D$$

then: $\quad k_1 \;=\; \dfrac{[C][D]}{[A][B]} \quad$ and $\quad k_2 \;=\; \dfrac{[A][B]}{[C][D]}$

Thus at equilibrium: $\quad k_1[A][B] \;=\; k_2[C][D]$

We conclude that at equilibrium, the equilibrium constant: $\quad K_E \;=\; \dfrac{k_1}{k_2}$

A Few Examples of Chelates

Anion	Cation	Complex
SO_4^{-2}	$\left[Cu(NH_3)_4\right]^{+2}$	$\left[Cu(NH_3)_4\right]SO_4$
$\left[CuBr_4\right]^{-2}$	$\left[(NH_4)_2\right]^{+2}$	$\left[CuBr_4\right](NH_4)_2$
Cl^{-1}	$\left[Co(NH_3)_6\right]^{+3}$	$\left[Co(NH_3)_6\right]Cl_3$
Cl^{-1}	$\left[Co(NH_3)_5Cl\right]^{+2}$	$\left[Co(NH_3)_5Cl\right]Cl_2$
Cl^{-1}	$\left[Co(NH_3)_4Cl_2\right]^{+1}$	$trans - \left[Co(NH_3)_4Cl_2\right]Cl$
Cl^{-1}	$\left[Co(NH_3)_4Cl_2\right]^{+1}$	$cis - \left[Co(NH_3)_4Cl_2\right]Cl$
Cl^{-1}	$\left[Be(OH_2)_4\right]^{+2}$	$\left[Be(OH_2)_4\right]Cl_2$

etc.

Thus, as an example, consider the following planar coordination complex.

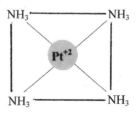

Organic Chemistry

Major Functional Groups

Chemical nomenclature is the chief objective here but, of course, the subject of organic chemistry has a beauty of its own. Understanding organic chemical structure depends upon reactive functional groups, and the following groups constitute many (but certainly not all) of the functional groups. Of course, a molecule may contain multiple functional groups.

Multiple Covalent Bonds

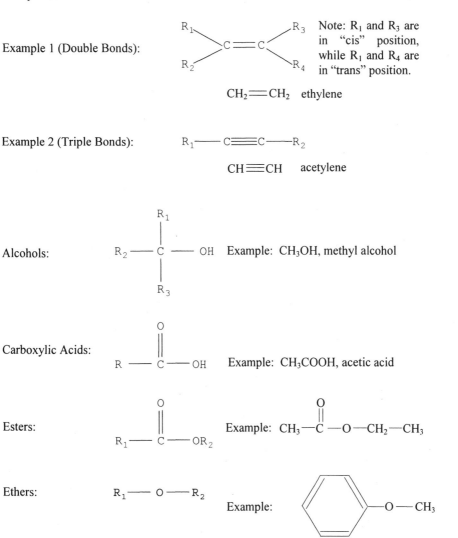

Example 1 (Double Bonds):

Note: R_1 and R_3 are in "cis" position, while R_1 and R_4 are in "trans" position.

CH_2==CH_2 ethylene

Example 2 (Triple Bonds): R_1——C≡≡C——R_2

CH≡≡CH acetylene

Alcohols:

Example: CH_3OH, methyl alcohol

Carboxylic Acids:

Example: CH_3COOH, acetic acid

Esters:

Example: CH_3—C—O—CH_2—CH_3

Ethers: R_1——O——R_2

Example:

Peroxides: 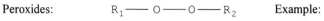 Example:

$$(CH_3)_3-C-O-O-(CH_3)_3$$

Aldehydes: Carbonyl group:

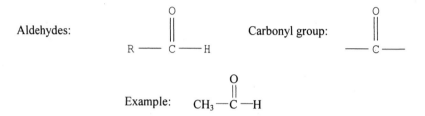

Ketones: Example:

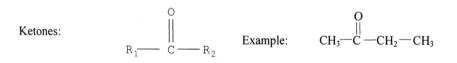

Aromatics:

benzene

Other examples:

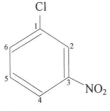

1-chloro-3-nitrobenzene

It is common to number carbon atoms, or other atoms in the case of heterocyclic molecules, so that it is clear to which atom reference is being made. There will be many cases in which this becomes important.

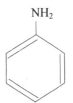

aminobenzene

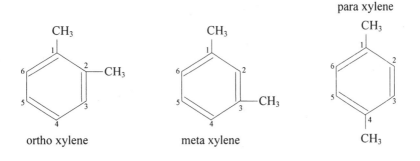

ortho xylene   meta xylene   para xylene

Heterocyclic aromatics (they contain atoms other than carbon, such as nitrogen, sulphur, etc. in a ring); we shall find many examples of these later on.

Structural Isomers and Tautomers

There are different kinds of isomers. Structural isomers are molecules with the same number of carbons, hydrogens, nitrogens, etc., but with different structures and different properties. Examples follow. Isomers are important because molecular rearrangements that provide increased stability can thus be understood. Isomers that easily interconvert explain different possible reactions. Thus upon examination, one might find that a chemical transformation is easily explained, when one realizes that a tautomeric form is closely related to an end product, while an entirely different end product is explained by another tautomeric form.

Structural Isomers of $C_5 H_{12}$ (An interesting computation is to determine for these alkanes, the number of structural isomers, given the number of carbons)

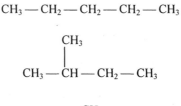

$$CH_3 — CH_2 — CH_2 — CH_2 — CH_3$$

$$CH_3 — \overset{\displaystyle CH_3}{\underset{\displaystyle |}{CH}} — CH_2 — CH_3$$

$$CH_3 — \overset{\displaystyle CH_3}{\underset{\displaystyle \underset{\displaystyle CH_3}{|}}{\overset{|}{C}}} — CH_3$$

Given "n" carbons, how many structural isomers are there?

Enantiomorphic Isomers

There are yet other kinds of isomer than those above. Another class of isomers is called enantiomorphs, which rotate polarized light at different angles and are also referred to as stereoisomers or mirror-image isomers. If a carbon atom is bonded to four different functional groups, then the molecule forms enantiomorphs.

$$R_1 — \overset{\displaystyle R_3}{\underset{\displaystyle R_4}{\overset{|}{\underset{|}{C}}}} — R_2$$

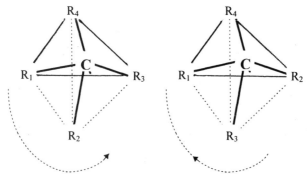

To show that the above carbon is optically active, it is common to write a star at the active carbon:

Note that a molecule may contain more than one optically active carbon. A carbon with a star denotes an optically active carbon. The following has $4 = 2^2$ enanteomorphic isomers.

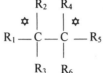

As a point of interest, silicon, with four different functional groups, has stereoisomers.

When a solution (containing one of the enantiomorphic molecules) is viewed in polarized light, the light is rotated left or right. Right rotation is referred to as dextro (D), while left rotation is referred to as levro (L). Amino acid forms found in living things are D-amino acids. Thus enantiomophic molecules are quite relevant in this book. See Figure 1.1.

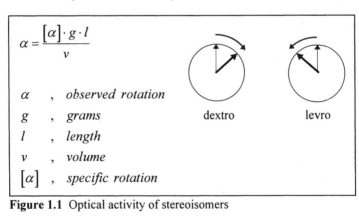

Figure 1.1 Optical activity of stereoisomers

Typically, the monochromatic D lines of Sodium, or the green line of mercury, or the red line of cadmium are used, at a fixed temperature ($25°C$).

Recall that tautomers are spontaneously interconvertible isomers. As an example:

$$R — CH_2 — CH = O \rightleftharpoons R — CH = CH — OH$$

Another example of tautomers are the lactim and lactam forms of uracil (a pyrimidine derivative).

Pyrimidine exists in lactim and lactam tautomers. As an example, uracil is a derivative of pyrimidine (see the section "Ribose, Deoxyribose, Phospho-Diesters" to get an example of the relevance of these tautomers).

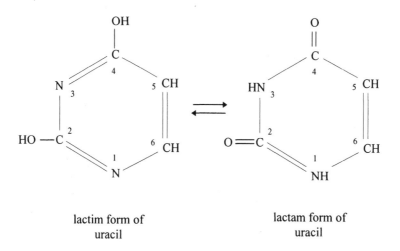

lactim form of
uracil

lactam form of
uracil

Tautomers

Similarly there are tautomeric forms derived from purine.

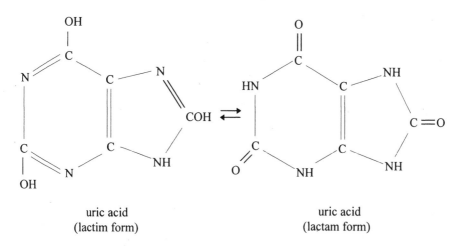

uric acid
(lactim form)

uric acid
(lactam form)

One should keep in mind that "stability" is significant, and stability enters consideration in a number of ways. One common way that stability should be considered is in molecular rearrangements; thus tertiary carbon is more stable than secondary carbon, which in turn is more stable than primary carbon. Thus, for example, a secondary group might rearrange into a tertiary group, and a surprising result may be found.

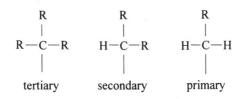

tertiary　　　　secondary　　　primary

Aldehydes, Ketones, and Saccharides

Sugars form the basis of many important biological macromolecules. A short examination of some sugar molecules will be very helpful. Sugars may be simple monosaccharides, or more complex saccharides. Oligosaccharides are sugars composed of a few saccharides, such as disaccharides. Polysaccharides are sugars composed of many saccharides. When polysaccharides are composed of the same sugar(s), they are referred to as homo-polysaccharides. To begin with, our major interest will be monosaccharides, then as homo-polysaccharides.

Saccharides derived from
aldehydes and ketones

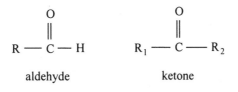

aldehyde　　　　　　　ketone

Monosaccharides are aldehydes or ketones with two or more hydroxyl groups. Monosaccharides are aldoses or ketoses. Although it will be omitted, at this point, stereoisomeric carbons will be starred (✿).

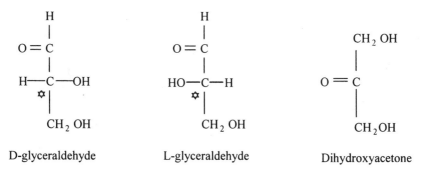

D-glyceraldehyde　　　　L-glyceraldehyde　　　　Dihydroxyacetone

Aldoses with 4, 5, 6, or 7 carbons are named tetroses, pentoses, hexoses, and heptoses.

Haworth Projections

Pentoses and hexoses cyclize to form furanose and pyranose rings. In order to understand both structure as well as nomenclature, a few notes follow.

For D-glucose drawn as Haworth projections, the designation α means that the hydroxyl group attached to the C-1 carbon is below the plane of the ring; β means that the hydroxyl group is above the plane of the ring. For D-fructose, The C-2 carbon is referred to as the anomeric carbon atom; thus the α and β forms are anomers.

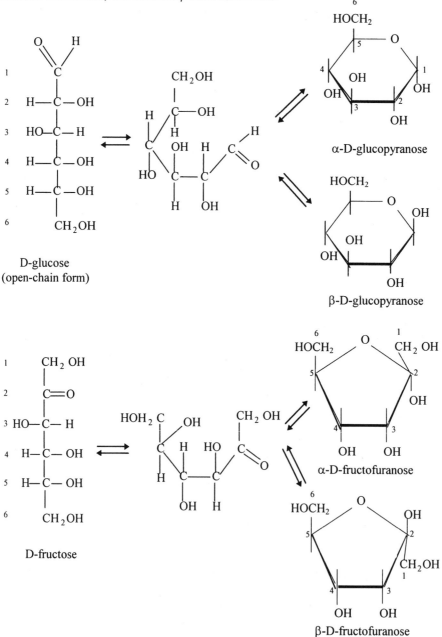

α-D-glucopyranose

β-D-glucopyranose

D-glucose
(open-chain form)

D-fructose

α-D-fructofuranose

β-D-fructofuranose

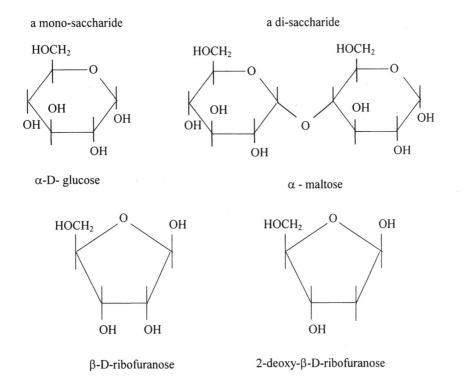

Two α-D-glucoses are
1-4 bonded as α - maltose

a mono-saccharide   a di-saccharide

α-D- glucose   α - maltose

β-D-ribofuranose   2-deoxy-β-D-ribofuranose

Biological Macromolecules

Molecules of glucose, ribose, and deoxyribose may be linked together to form polymeric chains or macro-molecules, commonly found in the study of biology. Glycogen is composed of α-1,4 linked glucose, and cellulose is composed of β-1,4 linked glucose.

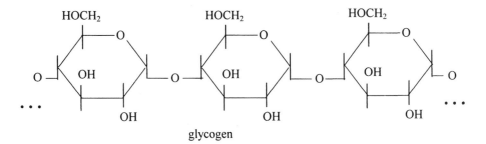

glycogen

Ribose, Deoxyribose, Phospho-Diesters

Ribose and deoxyribose carbons are numbered as follows.

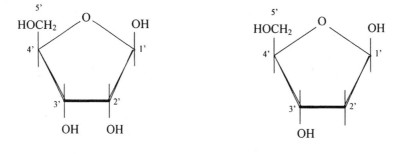

β-D- ribofuranose   2'-deoxy-β-D-ribofuranose

Phospho-diesters may link together ribose or deoxyribose chains, at the 5' and 3' carbons to create polymers.

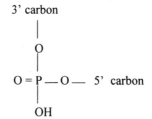

3' carbon

$O = P — O —$ 5' carbon

OH

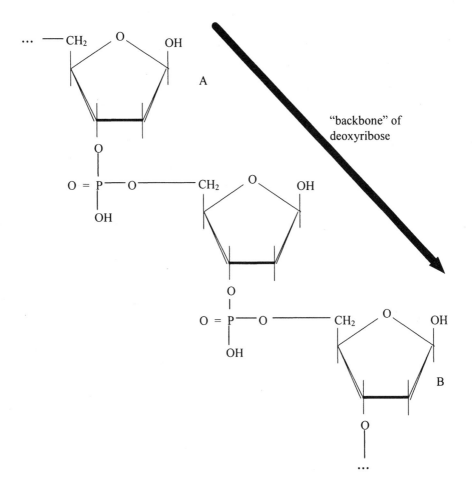

Note: One may refer to the chain from the 5' carbon at "A", to the 3' carbon at "B". This will become important when nucleotide bases are added to the DNA "backbone" to create DNA.

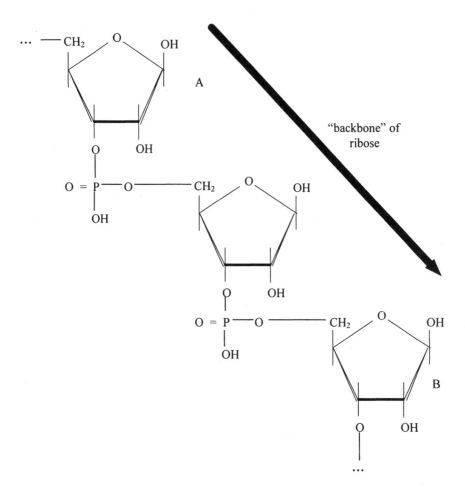

Note: One may refer to the chain from the 5' carbon at "A", to the 3' carbon at "B". This will become important when nucleotide bases are added to the RNA "backbone" to create RNA.

Pyrimidine Derivatives

Thymine, cytosine and uracil are pyrimidine derivatives, while adenine and guanine are purine derivatives. However, these derivatives are more easily understood if the lactim / lactam tautomers of pyrimidine and purine are examined.

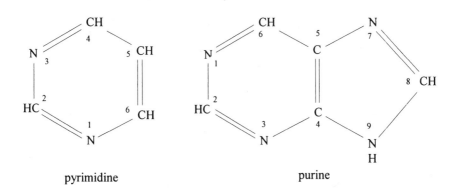

pyrimidine   purine

thymine

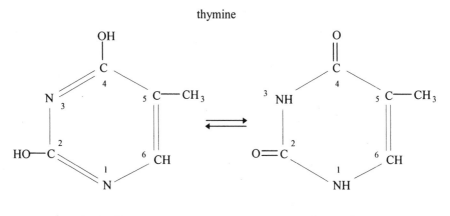

lactim form   lactam form

cytosine

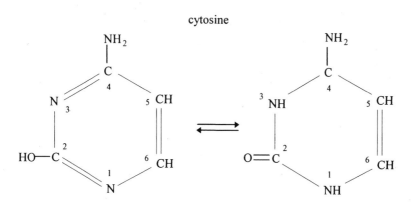

lactim form   lactam form

uracil

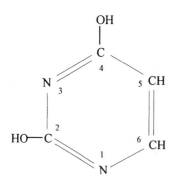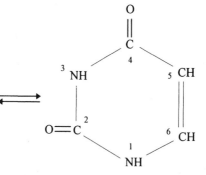

lactim form   lactam form

Purine Derivatives

adenine

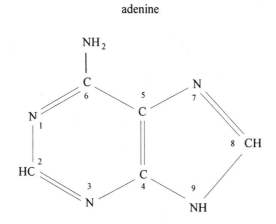

guanine

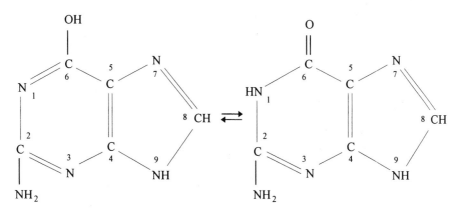

lactim form   lactam form

Nucleosides and Nucleotides

A nucleoside consists of a purine or pyrimidine base linked to a pentose.
A nucleotide consists of a phosphate ester of a nucleoside.

Nucleosides

Base	Ribonucleoside	Derivative
Adenine	Adenosine	Purine
Guanine	Guanosine	Purine
Uracil	Uridine	Pyrimidine
Cytosine	Cytodine	Pyrimidine

Base	Deoxyribonucleoside	Derivative
Adenine	Deoxyadenosine	Purine
Guanine	Deoxyguanosine	Purine
Thymine	Deoxythymidine	Pyrimidine
Cytosine	Deoxycytodine	Pyrimidine

Ribonucleosides

uridine

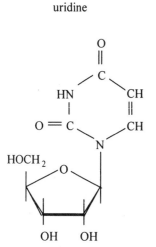

1-β-D ribofuranosyluracil

cytidine

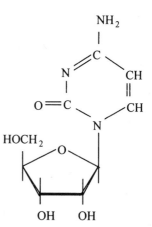

1-β-D ribofuranosylcytosine

adenosine

guanosine

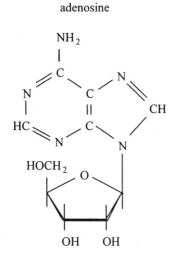

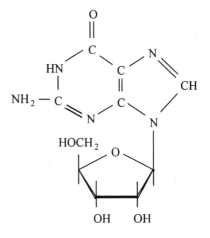

9-β-D ribofuranosyladenine

9-β-D ribofuranosylguanine

Deoxyribonucleosides

thymidine

cytidine

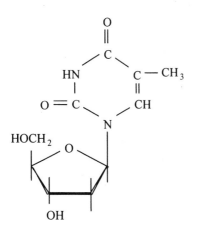

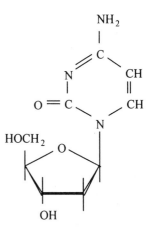

1-β-D deoxyribofuranosylthymine

1-β-Ddeoxyribofuranosylcytosine

adenosine

guanosine

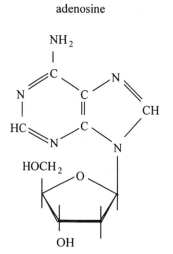

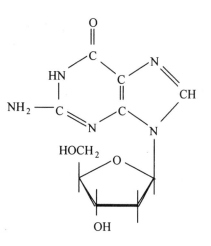

9-β-D deoxyribofuranosyladenine

9-β-D deoxyribofuranosylguanine

Nucleotides

Organic acids have the general formula RCOOH with the following structure.

Specific examples where there are different groups for R:

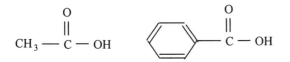

Esters have the general formula R_1COOR_2 with the following structure.

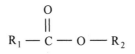

A specific example might be:

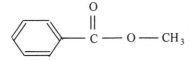

Nucleotides are phosphate esters of nucleosides: (linked to a nucleoside)

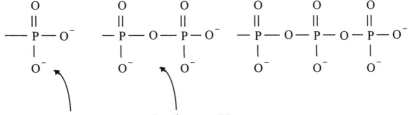

orthophosphate or P_i pyrophosphate or PP_i

Ribonucleotides

uridine 5'-phosphate

cytidine 3'-phosphate

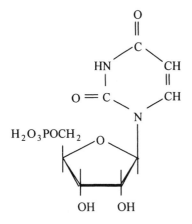

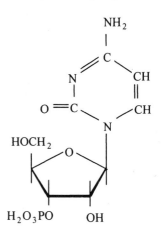

adenosine 5'-phosphate

guanosine 5'-phosphate

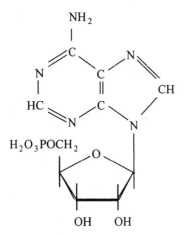

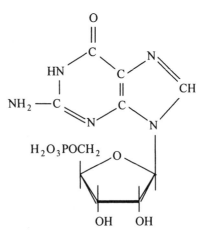

Deoxyribonucleotides

deoxythymidine 5'-phosphate

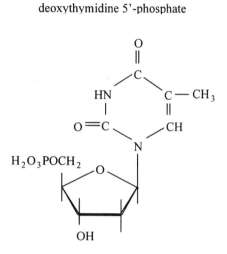

deoxycytidine 5'-phosphate

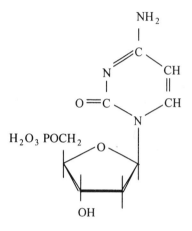

deoxyadenosine 5'-phosphate

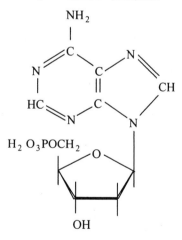

deoxyguanosine 5'-phosphate

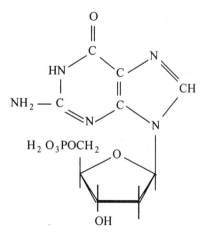

The nucleotides with phosphate groups linked to the 5' position are of specific interest.
DNA is a deoxyribose polymer linked to A, C, G, T nucleotides.
RNA is a ribose polymer linked to A, C, G, U nucleotides.

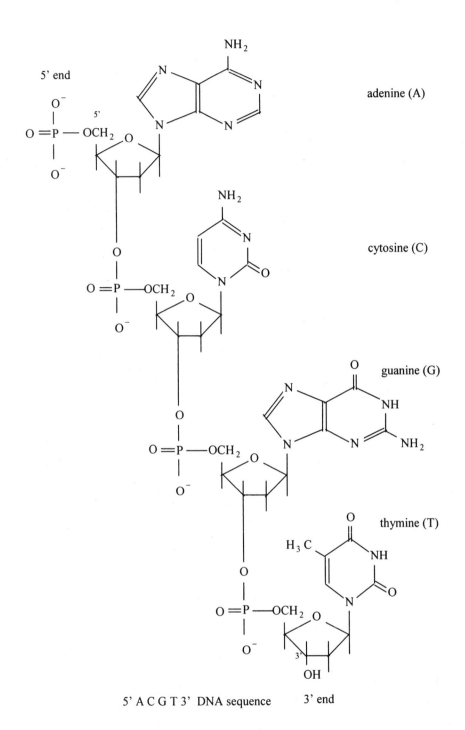

5' end

5'

adenine (A)

cytosine (C)

guanine (G)

thymine (T)

5' A C G T 3' DNA sequence   3' end

DNA as a Double Helix

From the viewpoint of energy, a double helix configuration offers great stability. This will be discussed when dealing with RNA folding, but to get an idea of the relationships, there is bonding between the base pairs, as well as steric considerations, as well as hydrogen bonding that confers stability. The DNA bases are hydrophobic and thus tend to be located internally with respect to the helix. Hydrogen bonds tend to be external to the helix, providing stability by bonding to the water in an aqueous environment.

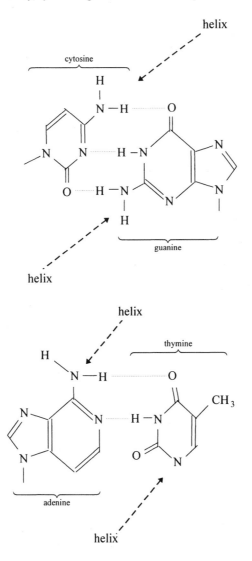

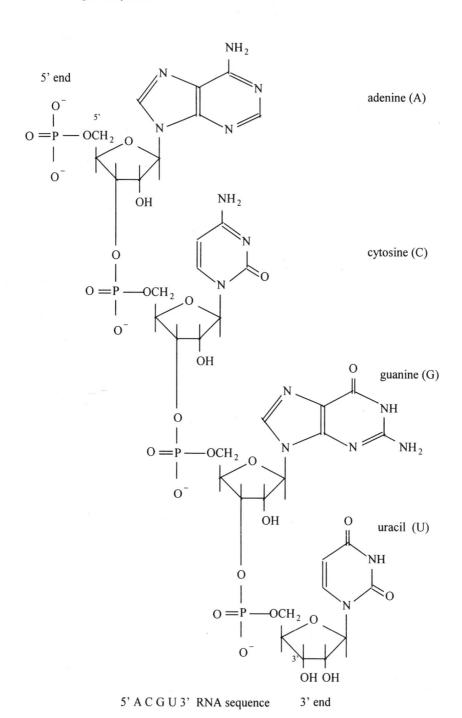

adenine (A)

cytosine (C)

guanine (G)

uracil (U)

5' A C G U 3' RNA sequence   3' end

Energy in Biological Systems

The phosphate linkage is a source of energy that is used to drive various biochemical reactions. The structures involved are as follows.

adenosine triphosphate or ATP

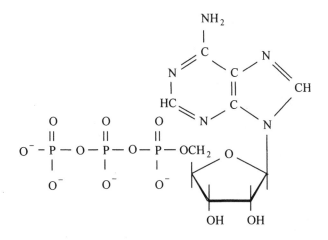

adenosine diphosphate or ADP

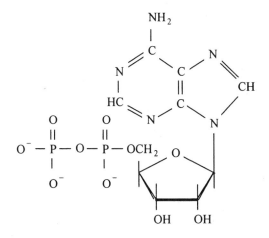

adenosine monophosphate or AMP

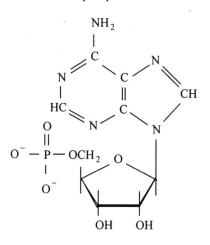

guanosine triphosphate or GTP

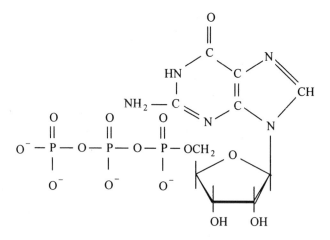

guanosine diphosphate or GDP

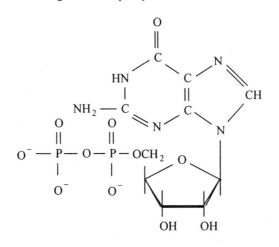

guanosine monophosphate or GMP

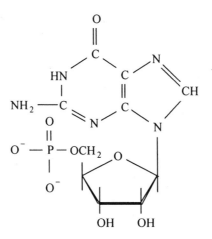

thymidine triphosphate or TTP

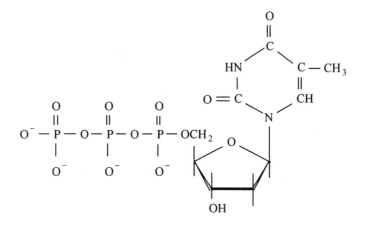

Note: this is a deoxyribose form.

thymidine diphosphate or TDP

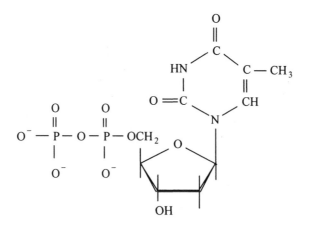

Note: this is a deoxyribose form.

thymidine monophosphate or TMP

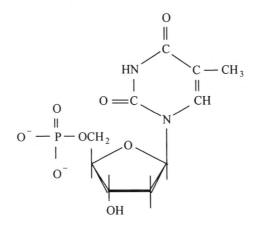

Note: this is a deoxyribose form.

cytidine triphosphate or CTP

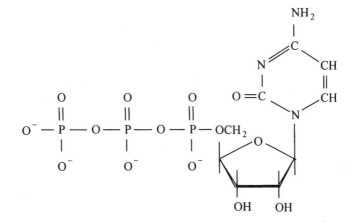

cytidine diphosphate or CDP

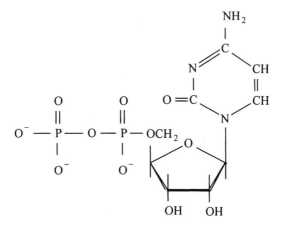

cytidine monophosphate or CMP

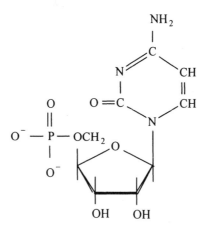

uridine triphosphate or UTP

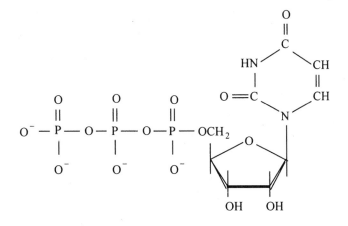

uridine diphosphate or UDP

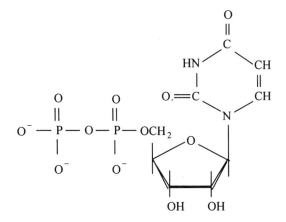

uridine monophosphate or UMP

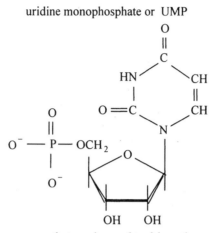

These nucleotides release energy that can be used to drive other reactions, thus:

$$ATP + H_2O \rightleftharpoons ADP + P_i + H^+ + energy$$

$$ATP + H_2O \rightleftharpoons AMP + PP_i + H^+ + energy$$

Analogous reactions occur as follows.

$$XTP + H_2O \rightleftharpoons XDP + P_i + H^+ + energy$$

$$XTP + H_2O \rightleftharpoons XMP + PP_i + H^+ + energy$$

where X may be C, T, G, or U.

ATP, ADP, and AMP are interconvertible.

$$ATP + AMP \rightleftharpoons 2\,ADP$$

and in general:

$$ATP + XMP \rightleftharpoons ADP + XDP$$

but also:

$$XTP + YDP \rightleftharpoons XDP + YTP$$

In general, all these forms are interconvertible. In addition to XTP, there are dXTP (deoxyadenosine, deoxyguanosine, deoxycytidine, deoxythymidine) molecules that not only enter into energy relationships, but simultaneously provide the bases needed for DNA replication [170, p. 740], [90, p. 90].

An example where ATP is used to drive a reaction, is the following, the first step in glycolysis.

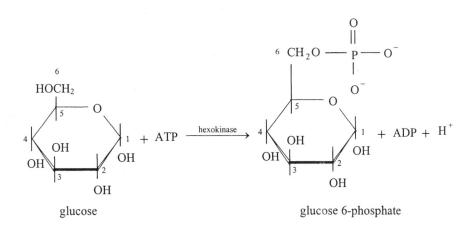

glucose   glucose 6-phosphate

Amino Acids and Proteins

Amino acids of interest in this book are alpha amino acids. Amino acids may be classified into different families, depending upon the carbon chain as follows.

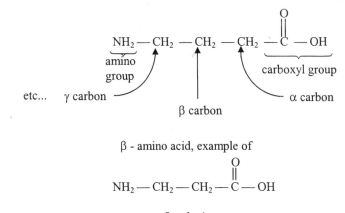

β - amino acid, example of

$$NH_2 - CH_2 - CH_2 - C - OH$$

β - alanine

Note that the α carbon for all α-amino acids (except for glycine) is optically active.

The 20 Common α Amino Acids Found in Proteins

$$NH_2 - CH_2 - \overset{\overset{\displaystyle O}{\|}}{C} - OH$$

glycine (gly)

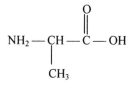

L - alanine (ala)

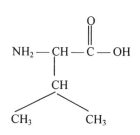

L - valine (val)

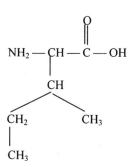

L - isoleucine (ileu)

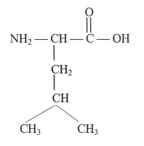

L - leucine (leu)

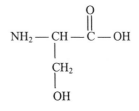

L - serine (ser)

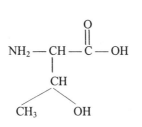

L - threonine (thr)

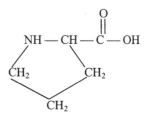

L - proline (pro)

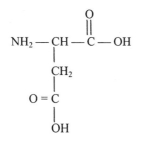

L - aspartic acid (asp)

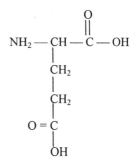

L - glutamic acid (glu)

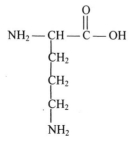

L - lysine (lys)

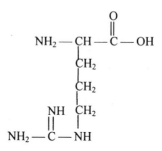

L - arginine (arg)

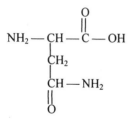

L - asparagine (asn)

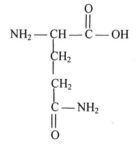

L - glutamine (gln)

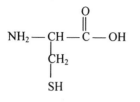

L - cysteine (cys)

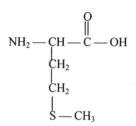

L - methionine (met)

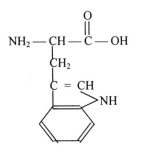

L - tryptophan (try)

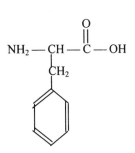

L - phenylalanine (phe)

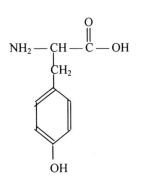

L - tyrosine (tyr)

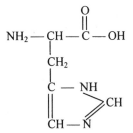

L - histidine (his)

Amino acid	3-Letter code	1-Letter code
alanine	Alu	A
arginine	Arg	R
aspartic acid	Asp	D
asparginine	Asn	N
cysteine	Cys	C
glutamic acid	Glu	E
glutamine	Gln	Q
glycine	Gly	G
histine	His	H
isoleucine	Ile	I
leucine	Leu	L
lysine	Lys	K
methionine	Met	M
phenylalanine	Phe	F
proline	Pro	P
serine	Ser	S
threonine	Thr	T
tryptophan	Trp	W
tyrosine	Tyr	Y
valine	Val	V

Figure 1.2 Amino acid abbreviations

The Central Dogma of Genetics

Self replication (see Figure 1.3)

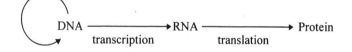

DNA —————→ RNA —————→ Protein
$\qquad$ transcription $\qquad\qquad$ translation

Watson-Crick Complementary Base-Pairs

DNA				RNA		
A	pairs with	T		A	pairs with	U
T	pairs with	A		U	pairs with	A
C	pairs with	G		C	pairs with	G
G	pairs with	C		G	pairs with	C

Note that there is a purine matched with a pyrimidine derivative in each pair of complementary bases. However, mismatches do occurr.

If $\overline{X}$ is the complement of X, then the Watson-Crick complements are as follows.

__DNA__			__RNA__	
$\overline{A}$	=	T	$\overline{A}$	= U
$\overline{T}$	=	A	$\overline{U}$	= A
$\overline{C}$	=	G	$\overline{C}$	= G
$\overline{G}$	=	C	$\overline{G}$	= C

DNA is composed of
words over the alphabet
A, C, G, T

RNA is composed of
words over the alphabet
A, C, G, U

DNA exists as a double helix [4], [30, p. 156], [87, p. 15] composed of bases A, C, G, T linked to a dexoyribose strand, coupled with another strand in which each base is matched by its complement.[a]

Example: 5' A C C T G A C 3'   strand 1
               3' T G G A C T G 5'   strand 2

RNA exists as a single strand, typically composed of bases A, C, G, U linked to a ribose strand.

Example: 5' A C C U G A C 3'

---

[a] Triple helix, quadruple, and quintuple forms have been found (and have value pharmaceutically).

Replication

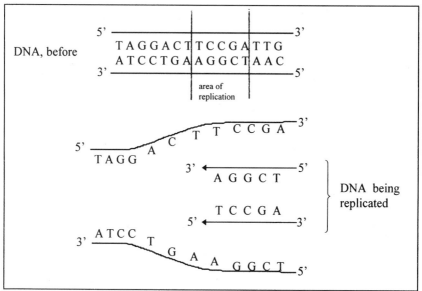

Figure 1.3 DNA replication

Transcription

Transcription means that a complementary copy of RNA is copied from DNA at an active site (where the double helix strands spread apart). See Figure 1.4.

Example:

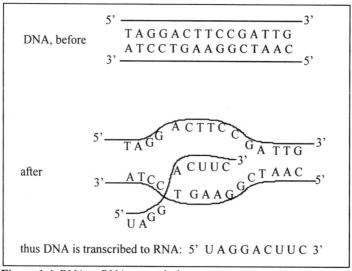

thus DNA is transcribed to RNA: 5' U A G G A C U U C 3'

Figure 1.4 DNA to RNA transcription

Genetic Code

It has been experimentally deduced that every three non-overlapping RNA bases (codons) uniquely specify an amino acid. Thus there are 64 possible triplets:

$$\left|\{A,C,G,U\}\right| \times \left|\{A,C,G,U\}\right| \times \left|\{A,C,G,U\}\right| = 4^3 = 64 .$$ The following table specifies the codons found.

1st	U		C		A		G	3rd
					2nd			
U	Phe		Ser		Tyr		Cys	U
U	Phe		Ser		Tyr		Cys	C
U	Leu		Ser		stop		stop	A
U	Leu		Ser		stop		Trp	G
C	Leu		Pro		His		Arg	U
C	Leu		Pro		His		Arg	C
C	Leu		Pro		Gln		Arg	A
C	Leu		Pro		Gln		Arg	G
A	Ile		Thr		Asn		Ser	U
A	Ile		Thr		Asn		Ser	C
A	Ile		Thr		Lys		Arg	A
A	Met		Thr		Lys		Arg	G
G	Val		Ala		Asp		Gly	U
G	Val		Ala		Asp		Gly	C
G	Val		Ala		Glu		Gly	A
G	Val		Ala		Glu		Gly	G

There is a great deal of redundancy; thus for example, Ser has codons UCU, UCC, UCA, UCG, AGU, and AGC. The three nonsense stop codons (nonsense, as these do not code for amino acids) are UAG (amber), UAA (ochre), and UGA (opal).

Thus, for example, given mRNA: UUUUCGAACUGGCCAGUUGUGG..., we obtain:

U UUU CGA ACU GGC CAG UUG UG(•••

Phe Arg Thr Gly His Leu Trp•••

thus the sequence of amino acids found in the mRNA encoding corresponds to:

Phe Arg Thr Gly His Leu Trp...

A gene must start with Met codon, and end with a stop codon. However, to actually create protein, a more complex process intervenes.

"Frames" refer to the codons. Is a codon every three bases? Given bases ABCDEFGHIJ..., the codon frame corresponds to ABC DEF GHI..., but if Q is an error in QBCDEFGHIJ..., then sometimes there is a frame error, to get BCD EFG HIJ..., and similarly, in AQCDEFGHIJ..., then sometimes there is a frame error to get CDE FGH, ...

Genes

E. coli will be used to describe genes. Several proteins called RNA polymerase bind to base patterns (called promoter sequences), then proceed down the DNA strands, transcribing the DNA into mRNA. Thus a gene might appear as follows.

5' ——— T T G A C A ——— T A T A A T ——— A U G| gene| stop ——— 3'

⎵⎵⎵⎵⎵⎵⎵⎵⎵⎵⎵⎵⎵⎵⎵⎵⎵⎵⎵⎵⎵⎵⎵⎵⎵⎵⎵⎵⎵
promotor sequence   Met

In fact, the situation is often more complex than the simplistic scheme above. Often a gene is partitioned into exons E_i, with what are often called "junk" introns I_j separating the exons. In such a situation, the mRNA assembles the gene, excising the introns.

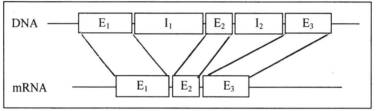

Figure 1.5 Gene intron, extron and mRNA

Operons

It has been pointed out that "First, not all DNA sequences (of bases) code for protein; many sequences contain other types of information, for example, regulatory signals controlling the synthesis of proteins. Secondly, the capacity for information storage appears to involve more than the one-dimensional pattern of bases; for example, dynamic information is stored in the three-dimensional conformation of DNA." It has also been stated that "Thus, in addition to coding for protein sequences, the DNA must code for its involvement with the cellular machinery. This includes instructions concerning the regulation and execution of protein synthesis, as well as instructions for the packaging, storing, and manipulation of DNA within the cell" [27, pp. 159-161].

We already know that transcription refers to mRNA formation from DNA, and we already know that the creation of proteins from mRNA codons is called translation. We must also be aware that transcription involves: recognition, binding, initiation, and termination sites all referred to as an "operon." The description of the following Lac operon has also been provided [27, pp. 159-161].

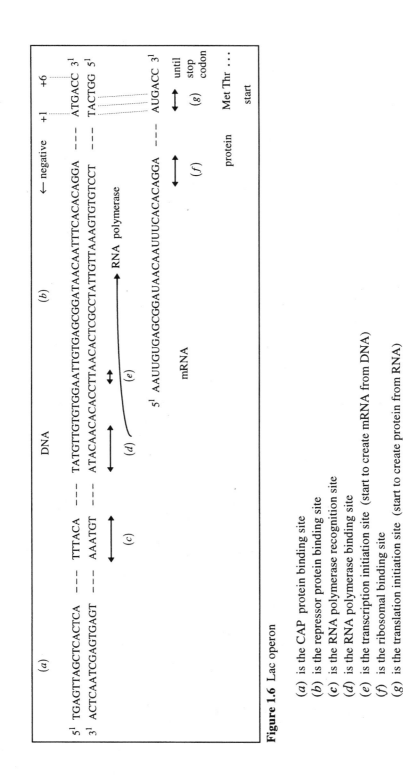

Figure 1.6 Lac operon

(a) is the CAP protein binding site
(b) is the repressor protein binding site
(c) is the RNA polymerase recognition site
(d) is the RNA polymerase binding site
(e) is the transcription initiation site (start to create mRNA from DNA)
(f) is the ribosomal binding site
(g) is the translation initiation site (start to create protein from RNA)

Note that both binding sites, the CAP protein binding site (a) and the repressor binding site (b), are palindromic (using Watson–Crick complements).

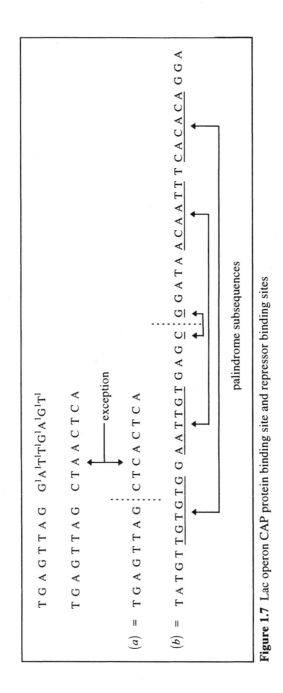

Figure 1.7 Lac operon CAP protein binding site and repressor binding sites

The way the operon functions:

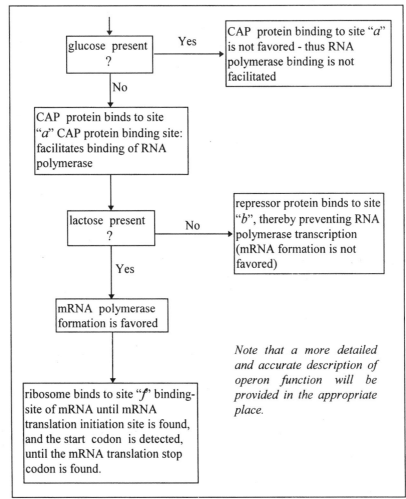

Figure 1.8 Operon function

Protein Assembly

Given that mRNA and amino acids are present, the question arises as to how proteins are assembled. It is posited that tRNA, a spatial adapter molecule of about 80 bases is linked to amino acids, and in the presence of mRNA codons, creates proteins.

RNA is single stranded, and energetic stability is enhanced by the RNA strand folding back upon itself, into helical regions.

Example:

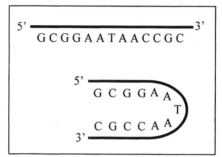

Figure 1.9 RNA folding

RNA Folding

Types of RNA folding: Primary, Secondary, Tertiary

Primary folding refers to the linear order of bases.
Secondary folding consists of planar relationships between bases, and there is a variety of folds possible.

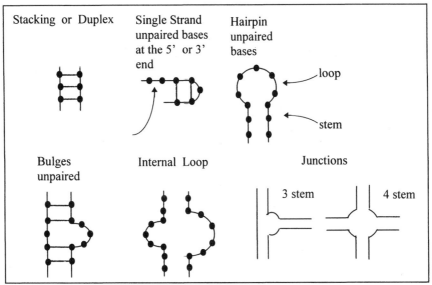

Figure 1.10 Names of kinds of RNA folds

Tertiary folding; there is a variety:

    Base Pair

        Pseudoknots: Unpaired bases in the secondary structure link to form folds.

    Single Strand:

        Base Triples (a base is simultaneously linked to two bases):

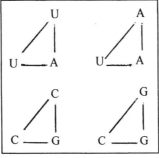

Figure 1.11 Base triples

Helix-Helix: The example is a triple helix.

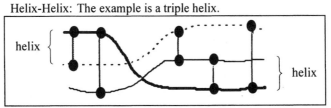

Figure 1.12 Triple helix

A number of tRNA sequences have been determined, and all have a cloverleaf pattern.

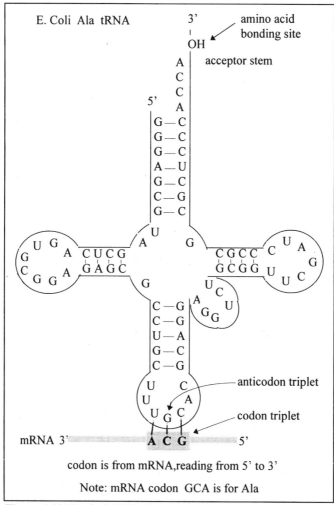

Figure 1.13 Typical tRNA cloverleaf

Inosine may appear in the anticodon. In addition, the anticodon does not necessarily correspond uniquely to a specific codon due to steric effects, referred to as "wobble" [170, pp. 886, 887].

mRNA interacts with tRNA to make proteins at cellular sites referred to as ribosomes. There may be multiple forms of equivalently folded tRNA, and clearly, all the amino acids found in proteins require a tRNA "adapter". Thus for example, we may find such tRNA adaptors as the ones on the next page [94, pp. 171-174].

tRNA, mRNA

$tRNA_1^{Leu}$, $tRNA_2^{Leu}$, $tRNA^{Asp}$, $mutRNA_{CUA}^{Tyr}$, or $m-tRNA_{CUA}^{Tyr}$ the last two are especially interesting, as they mean: a mutant form of $tRNA_{CAU}^{Tyr}$, where the tRNA is normally charged with (carries) tyrosine (to be emplaced in a protein), where the tRNA is associated with the CUA anti-codon. The CUA anti-codon implies the codon is the complement of AUC, which is UAG (the "amber" nonsense stop codon). Similarly, $wt-tRNA_{GUA}^{Tyr}$ refers to the wild-type tRNA for tyrosine, with anti-codon GUA, where the codon that corresponds to GUA is the complement of AUG, or UAC (the codon for tyrosine). Unfortunately, alternative notations are used that appear quite similar, but mean something very different; thus the literature must be examined to determine the context. We present a few examples from the literature. These are more useful in that the specific molecule that is bonded to tRNA is explicitly indicated.

$O-methyl-L-tyrosyl-tRNA_{CUA}^{Gly}-dCA$

$L-3-iodotyrosyl-tRNA_{CUA}^{Gly}-dCA$

$L-phenyllactyl-tRNA_{CUA}^{Gly}-dCA$

$N-Methyl-L-phenylalanyl-tRNA_{CUA}^{Gly}-dCA$

$5-aminovaleryl-tRNA_{CUA}^{Gly}-dCA$

$L-phenylglycyl-tRNA_{CUA}^{Gly}-dCA$

$L-2-amino-3,3-dimethylbutyryl-tRNA_{CUA}^{Gly}-dCA$

$2-amino-4-phosphonobutyryl-tRNA_{CUA}^{Gly}-dCA$

$\beta-phenylalanyl-tRNA^{Phe}$

$phenylglycyl-tRNA^{Phe}$

$I-Tyr-tRNA_{CUA}^{Gly}$

$glycyl-tRNA_{CUA}^{Gly}$

$Abu-tRNA^{Val}$

Aminoacyl-tRNA Synthetases

Sometimes the specific bacterial strain may be specified. Thus the tyrosyl-tRNA synthetase $m-TyrRS_{Mj}$ is from a mutant Methanococcus jannaschii, and the tyrosyl-tRNA synthetase $wt-TyrRS_{Ec}$ is a mutant of Escherichia coli. This will be discussed in greater detail in Chapter 3.

Peptide Bond and Proteins

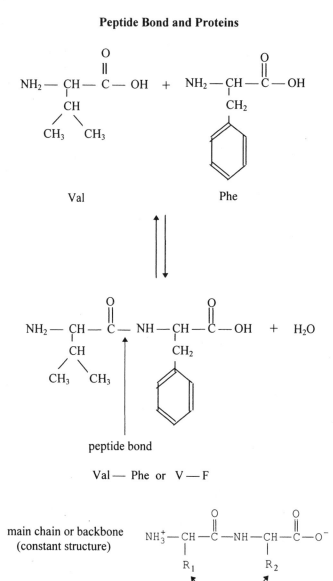

peptide bond

Val — Phe or V — F

main chain or backbone
(constant structure)

variable side chains

Protein chains may contain 20 to 2000 amino acid residues, and as amino acid residues have a mean molecular weight of 110, then the molecular weight is 2,200 to 220,000 (written 2.2 kd to 220 kd). Proteins typically have a complex folded pattern. As cysteine and methionine contain sulphur, disulphide bonds may act as bridges between side chains on the same or multiple backbones.

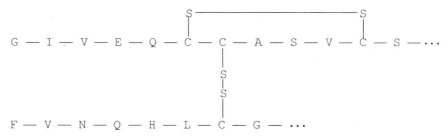

Proteins have a three-dimensional conformation that confers biological activity. Polypeptide conformations were examined crystallographically, and two structures were proposed:

α helix   (tightly coiled, right-handed sense)

β pleated sheets

The α helix exists as single and double strands (and multiplicities beyond two) due to NH–CO hydrogen backbone bonding. β pleated sheets have NH–CO hydrogen bonding between different backbones and the strands may be in the same (parallel) direction or opposite (anti-parallel) direction. Such anti-parallel structures are called β turns.

Primary structure:	sequence of amino acids
Secondary structure:	spatial arrangement of nearby amino acid residues (steric considerations, periodic structure)
Tertiary structure:	spatial arrangement of amino acid residues that are far apart or due to disulphide bonds
Quaternary structure:	spatial arrangement of multiple polypeptide chains

Super secondary structure: clusters of secondary structures

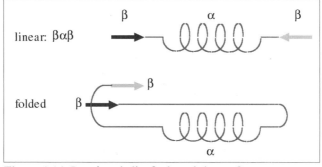

Figure 1.14 Protein α-helix, β pleated sheets, β turns

Repressor and activator proteins (for example, in operons) bind to DNA, but how? The three-dimensional conformations of α helices neatly fit into the DNA double helix (as a unit of recognition). We will not pursue this further, however [170, p. 962].

Transformations of DNA and RNA

Polymerases are enzymes that promote polymerization. Specifically, given a double-stranded DNA helix, the polymerization takes place respecting the Watson-Crick complements. As the OH (alcohol) functional group is quite reactive, extension takes place at the 3^1 end:

Example:

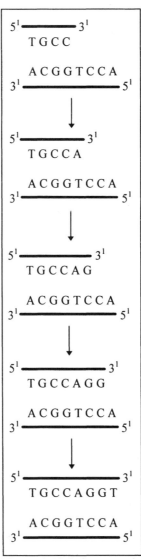

Figure 1.15 Polymerase dynamics

However, terminal transferase allows polymerization at 3^1 with single-stranded DNA.

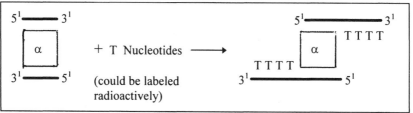

Figure 1.16 Terminal transferase polymerization

Whereas polymerases extend DNA, nucleases shorten DNA. There are two types of nucleases: endonucleases and exonucleases. Exonucleases remove nucleotides one at a time, from the 3^1 or 5^1 ends, and some exonucleases operate on either single- or double-stranded DNA.

Notation: restriction enzymes are classified as follows.
vwxyz where vwx names an organism, y is a strain and is optional, and z is a Roman numeral (optional) to indicate different restriction enzymes from the same organism.

Examples: HaeII, HaeIII, BamHI, EcoRI

Examples:
(In these examples, N and N^1 refer to any nucleotides, where N and N^1 are Watson-Crick complements. If there is no complementary base, then only N will be utilized.)

Exonuclease III removes nucleotides from 3^1 on double-stranded DNA, while Bal31 cleaves nucleotides from both 3^1 and 5^1 ends of double-stranded DNA simultaneously.

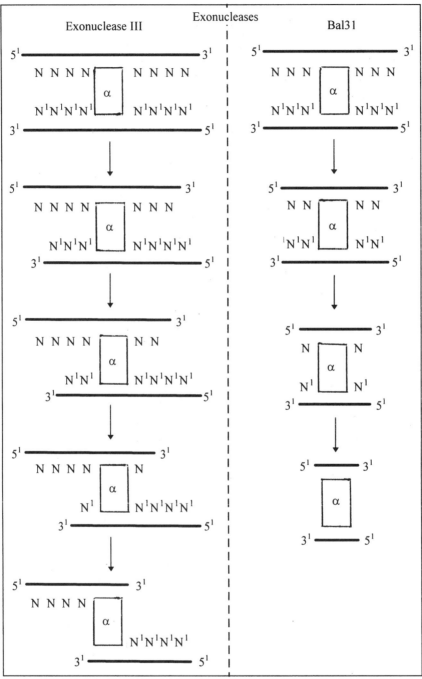

Figure 1.17 Exonuclease dynamics

Endonucleases break internal phospho-diester bonds; they cut either single or double strands of DNA internally.

Examples: Endonuclease S1 cuts single strands at any internal location.
            Endonuclease DNaseI cuts double strands at any internal location.

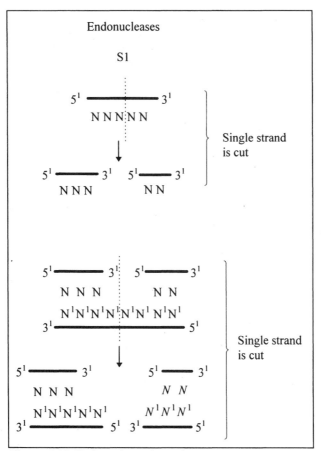

Figure 1.18 Endonuclease dynamics

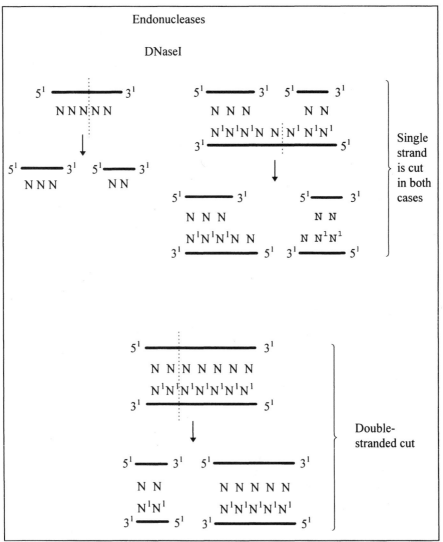

Figure 1.19 Dnase I endonuclease dynamics

Restriction Nucleases and Ligases

Restriction enzymes or restriction endonucleases also cleave DNA. Two uses for restriction enzymes are to deactivate foreign invading DNA, and also to repair DNA.

A palindrome is a word that is spelled the same way reading left to right as well as right to left. For example, a palindrome over the letters x, y, and z is: xxzy yzxx. We shall deal in some depth with palindromes later, but at this point, DNA palindromes appear in the coupled helix of DNA, on opposite strands. In the following example, the letters on top read the same way in reverse on the complementary strand.

Figure 1.20 DNA double-stranded helix palindromic structure

Two examples of restriction enzyme cleavage follow:

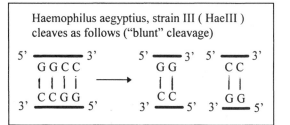

Figure 1.21 HaeIII restriction endonuclease blunt cleavage

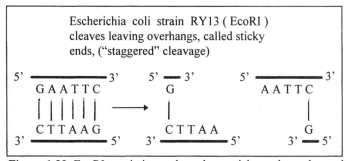

Figure 1.22 EcoRI restriction endonuclease sticky-end, overhang cleavage

Restriction endonuclease HgaI results in a staggered cut, but requires the recognition of 5^1 — GACGC as follows.

Restriction endonucleases require a recognition site before they can cleave. The cleavage may occur outside the recognition sequence (then it is called a Type I restriction endonuclease), or the cleavage may occur within the recognition sequence (in which case it is called a Type II restriction endonuclease).

Endonucleases are used naturally to chop up foreign DNA (from a virus, for example). The native DNA is protected by methylase which methylates native recognition sites so that endonucleases do not cleave the native DNA.

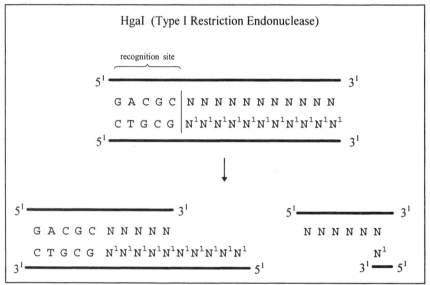

Figure 1.23 HgaI restriction endonuclease with recognition site and cleavage pattern

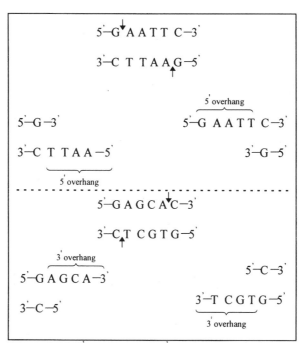

Figure 1.24 5' overhang and 3' overhang terminology

Ligases

Ligases are enzymes that reestablish phospho-diester bonds at $3^1 / 5^1$ and $5^1 / 3^1$ sites. In fact, ATP and NAD^+ have been found to act as ligases. It has been found that during ligation in vitro, hybridizations are possible, as the following example illustrates.

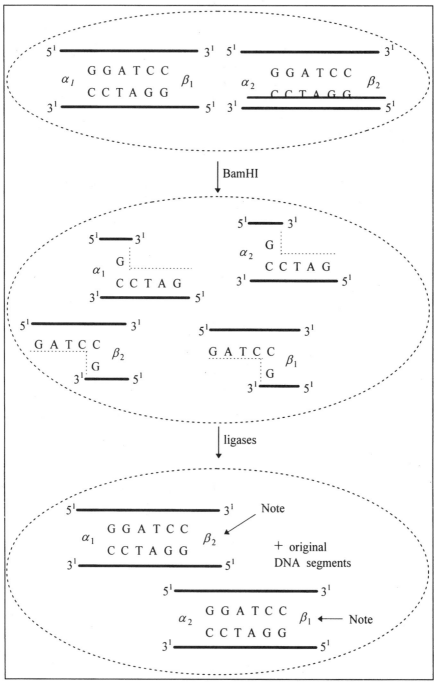

Figure 1.25 Ligase dynamics

Plasmids

While DNA are commonly thought of as linear, examination of some bacteria and protozoa has resulted in the discovery of plasmids. Plasmids are DNA material that may be either circular or linear, depending upon specific stages of their life cycle. In addition, mitochondria may contain circular DNA also. Thus sections of DNA may be excised or hybridized not only in linear sections of DNA, but even in circular sections of DNA. Could circular segments of DNA be recombined in ways that have mathematically interesting properties, or to make molecules with novel chemical properties? [b]

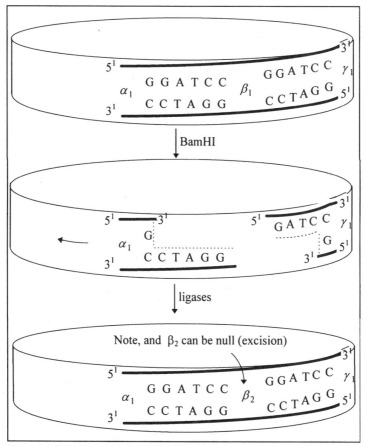

Figure 1.26 Plasmid (circular) DNA cleavage and ligase activity

---

[b] [94, p. 640] "The type I topoisomerases also can pass one segment of a single-stranded DNA through another. This **single-strand passage** reaction can introduce **knots** in DNA and can **catenate** two circular molecules so that they are connected like links on a chain. (Emphasis by B. Lewin, also [192]). Buckminsterfullerenes have similar structures. Applications of algebraic topology, graph theory, and knot theory!

Figure 1.27 illustrates the possibilities of plasmids recombining to get double or triple (or even larger) multiple-sized plasmids (see a and b). Molecular weight calculations using radioactive cesium as well as gel electrophoreses chromatography have found larger plasmids. Recombination into intertwined links (see d), or Möbius strips (see c) could also occur, as well as other mathematically interesting possibilities.

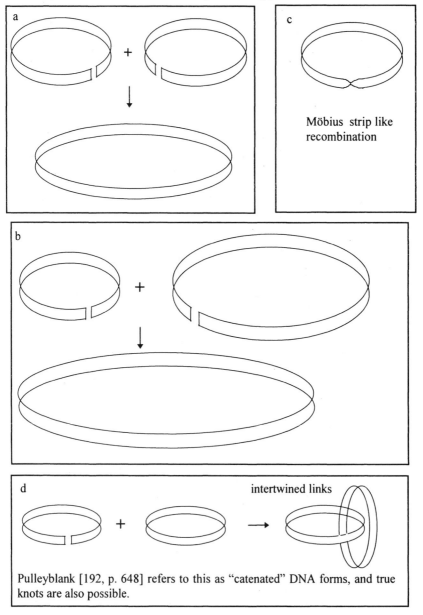

Möbius strip like recombination

Pulleyblank [192, p. 648] refers to this as "catenated" DNA forms, and true knots are also possible.

Figure 1.27 Possible plasmid (circular DNA) ligase topologies

In fact, to accurately characterize circular DNA, a triple is required, called the linking number (Lk), the twisting number (Tw), and the writhing number (Wr). These three numbers as a triple $< Lk, Tw, Wr >$ describe a topology [170, pp. 794-796]. Supercoiling may take place.

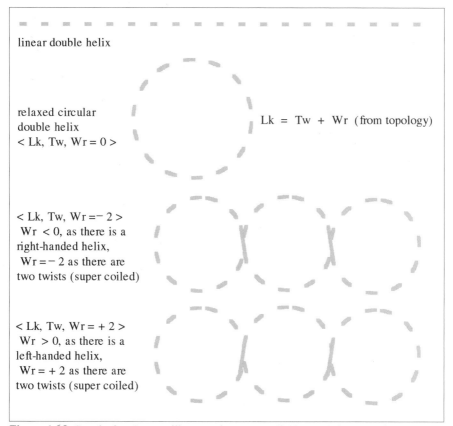

linear double helix

relaxed circular
double helix
$< Lk, Tw, Wr = 0 >$

$Lk = Tw + Wr$ (from topology)

$< Lk, Tw, Wr = -2 >$
$Wr < 0$, as there is a
right-handed helix,
$Wr = -2$ as there are
two twists (super coiled)

$< Lk, Tw, Wr = +2 >$
$Wr > 0$, as there is a
left-handed helix,
$Wr = +2$ as there are
two twists (super coiled)

Figure 1.28 Topologies: Supercoiling, topoisomerases, linking, twisting, writhing numbers

Topoisomerases, Supercoiling

Topoisomerases are enzymes that regulate the formation of these superhelical structures. Topoisomerases I make a transient break in only one strand of DNA supercoiled structures, then change the linking number by relaxing the supercoiling by reducing the linking number by one, then ligate or rejoin the broken strand. Topoisomerases II (a special subset being the gyrases) bind to supercoiled DNA and convert a circular piece of DNA into two supercoiled loops. Topoisomerases occur only in bacteria and are now being targeted in anti-cancer treatments, as well as other diseases such as Hodgkin's disease. Further details may be found in [47, pp. 58-65].

Selected Restriction Enzymes

While there are hundreds restriction enzymes, some of the more important ones are:

Microorganism	Restriction Enzyme	Restriction Site
Bacillus amyloliquefaciens H	BamHI	G↓G A T C C C C T A G↑G
Brevibacterium albidum	BalI	T G G↓C C A A C C↑G G T
Escherichia coli RY13	EcoRI	G↓A A T T C C T T A A↑G
Haemophilus aegyptius	HaeII	Pu G C G C↓ Py Py↑C G G C Pu
Haemophilus aegyptius	HaeIII	G G↓C C C C↑G G
Haemophilus influenzae R_d	HindII	G T Py↓Pu A C C A Pu↑Py T G
Haemophilus influenzae R_d	HindIII	A↓A G C T T T T C G A↑A
Haemophilus parainfluenzae	HpaI	G T T↓A A C C A A↑T T G
Haemophilus parainfluenzae	HpaII	C↓C G G G G C↑C
Providencia stuartii 164	PstI	C T G C A↓G G↑A C G T C
Streptomyces albus G	SalI	G↓T C G A C C A G C T↑G

Figure 1.29 Selected restriction endonuclease cleavage patterns

Microorganism	Restriction Enzyme	Restriction Site
Serratia marcescens	SmaI	C C C'G G G G G G‚C C C
Xanthomonas malvacearum	XmaI	C'C C G G G G G G C C‚C
Acetobacter aceti	AatII	G A C G T'C C‚T G C A G
Acinetobacter calcoaceticus 65	Acc65I	G'G T A C C C C A T G‚G
Bacillus caldolyticus	BclI	T'G A T C A A C T A G‚T
Deinococcus radiophilus	DraI	T T T'A A A A A A‚T T T
Moraxella bovis	MboI	G A G'C G G C T C‚G C C
Zoogloea ramigera	ZraI	G A C'G T C C T G‚C A G
Herpetosiphon giganteus	HgiAI	G W C C W'C W is A or T, thus four: G'A G C A C one of 4 C T C G T‚G

Many other restriction endonucleases, easily found on the Internet.
Most have different recognition sites; for example, Eco4III has 2 sites in Lambda,
4 sites in pBR322 (Boyer, with Resistance),[*] 2 sites in pACYC177
(Annie Chang).[*]

[*] These are plasmids constructed from components of other plasmids (courtesy, E. Lederberg).

Figure 1.30 Additional selected restriction endonuclease cleavage patterns

DNA/RNA Hybrid Cleavage

A question now presents itself. As RNA is so similar to DNA, is cleavage possible with RNA? One way to approach this question is to ask a few more questions. Restriction endonucleases appear to do their work on double-stranded helices. Second, what about the replacement of T by U in RNA (possibly a mechanism to prevent cleavage)?

The first question concerning double helices may be approached from at least two viewpoints:
1. RNA tends to coil back upon itself, assuming a double helix conformation that looks like the following:

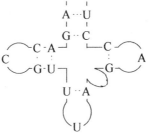

Effectively, it is like a DNA double helix. If steric considerations do not exclude endonuclease cleavage, then cleavage might be possible.
2. Does the existence of U in place of T prevent cleavage? Even if true, there can be substrings composed only of G and C bases that are subject to endonuclease cleavage.

The second question concerning the similarity of folded RNA to double helices of DNA can be broadened to hybrid DNA and RNA strands. Rather than continuing along these lines of thought, an interesting paper provides an affirmative answer concerning RNA cleavage [110]. To be explicit, HIV-RT (reverse transcriptase) RNase H functions as follows.

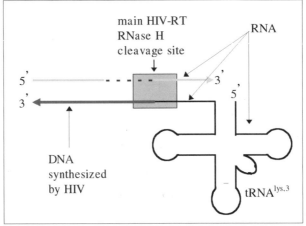

Figure 1.31 tRNA/DNA hybrid cleavage site

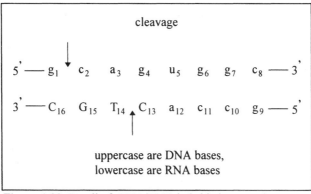

Figure 1.32 Detail of tRNA/DNA hybrid cleavage site

Introns in mRNA precursors are removed by splicing. For example, such splice sites are often recognized, as they begin with G U and end with A G. After the splice has removed the intron, the (now) adjacent exons are linked together. Small RNA molecules in the nucleus (snRNA) or in the cytosol (scRNA) form protein complexes (snRNP or scRNP) to effect this splicing. In addition, RNA can act as a self-splicing enzyme. We shall not pursue RNA splicing any further, however [170, pp. 862-865].

Chapter 2: A Review of Aspects of Automata Theory

Flagellants

"Flagellants proceeded in bands of 50 to 300; they moved in long, snakelike processions, two by two, in groups of a few hundred. The bands of flagellants walked with men in the front and women in the rear, chanting hymns. They dressed in cowled white robes emblazoned with red crosses on the front and back, and some carried crosses as well. Each band's leader was called 'master' or 'father.' He heard confessions and, to the horror of the clergy, imposed penance and granted absolution. Each member swore absolute obedience to the master for the duration of the procession, usually 33 1/3 days, which symbolized Christ's years on earth. The flagellants could not bathe, shave, or change their clothing; they could not sleep in soft beds and, although they were permitted to wash their hands once a day, it had to be done in a kneeling position, as a demonstration of humility. There were still more restrictions. Flagellants were forbidden to speak, even to one another, without the permission of the master. Sex was proscribed, and any male flagellant who said even a word to a woman had to kneel before the master and do penance. The master then beat him, chanting all the while, 'Arise by the honor of pure martyrdom and henceforth guard yourself against sin.'

"When flagellants came into a town or village, they made their way to the most prominent local church. There they formed a circle. The men took off their outer clothing and put on loose skirts, which fell from their waists to their feet. Next, they began their standard rite. The penitent flagellants marched around in a circle, took a crucifix position, and were scourged. Sometimes the penitents would flagellate themselves, singing hymns, celebrating Christ's passion and the glories of Virgin Mary. Generally, the master and two assistants stood in the center of the circle supervising the process and making sure that no one slackened in their enthusiasm. Three times during the rite all would fall down 'as though struck by lightening' and lie prostrate, sobbing. The master would walk among them, asking God for mercy on all sinners. Then the flagellism would continue.

"Much of this reflected the general dissatisfaction with the clergy, who were seen as corrupt and incapable of assuaging the pain of the Black Death in any way.

"By early 1349, the recessionals were dominated by marginal elements, including increasing numbers of vagabonds and criminals.

"For Pope Clement VI was fully informed concerning this fatuous new rite by the masters of Paris through emissaries recently sent to him, and, on grounds that it had been damnably formed, contrary to law, he forbade the flagellants under threat of anathema to practice in the future the public penance which they had so presumptuously undertaken" [227, pp. 70-72].

Review of Aspects of Automata Theory

A detailed study of automata theory, often also called the theory of computability, or albeit the more limited; mathematical linguistics, is not the purpose of this book. The purpose of this book is the relationship between automata theory and emergent computation, with special emphasis upon bioinformatics. Thus for a detailed review of a large part of automata theory usually covered in an introductory course, the reader should consult any number of books that cover this subject which also discus semigroups. A suggested reference is *Automata Theory*, by M. Simon [166].

As a brief summary of most of the important ideas, a relevant review of the subject of automata theory, is provided as an aid.

Sequential Machines

A sequential machine $M = \left(S,\ I,\ O,\ \delta,\ \lambda\right)$ where

S   is a finite set of states
I   is a finite set of inputs
O   is a finite set of outputs
δ is a single-valued function that maps the current state and input to the next state, or $\delta: S \times I \rightarrow S$

λ is a single-valued function, and sequential machines come in two equivalent flavors, Moore or Mealy machines, defined as follows.

Moore: $\lambda: S \rightarrow O$, λ maps the current state into an output.
Mealy: $\lambda: S \times I \rightarrow O$, λ maps the current state and input into an output.

Almost all references in this book from here on will use the symbol λ to refer to an identity or monoid element, as in a mathematical group or mathematical semigroup. Only when referring to a sequential machine will the symbol λ refer to a function. This also means that λ will refer to a word of length zero, as used in standard mathematical linguistics.

Semigroups and Monoids

Associated with each sequential machine, there is a semigroup, which if it isn't a monoid, may have a two-sided identity adjoined to it, to create a monoid. In this book, there is only a small discussion concerning semigroups or monoids, so little will be said about this subject.

Definition: A semigroup S is a collection of elements with one closed operation which is associative. There need not be a (two-sided) identity, but if there is a two-sided identity it is unique, and then the semigroup S is called a monoid. If an element x is idempotent, then $x^2 = x$. E(S) is the set of idempotents of S (and can be empty). We study semigroups no further in this book; an examination of [166] to gain further knowledge is recommended.

Linear Sequential Machines

A special subfamily of sequential machines is referred to as linear sequential machines (over a Galois Field), or an LSM. As with sequential machines already defined above, LSMs come in two flavors, Moore or Mealy LSMs. The difference is that the next-state function and the output function are defined over matrices and vectors, as follows.

$$
\begin{aligned}
Y_t &= A \cdot y_{t+1} + B \cdot x_{t+1} \quad \left(\text{next state}\right) \\
z_t &= C \cdot y_{t+1} \quad\qquad\quad \left(\text{output}\right)
\end{aligned}
\qquad \text{Moore}
$$

or

$$
\begin{aligned}
Y_t &= A \cdot y_{t+1} + B \cdot x_{t+1} \quad \left(\text{next state}\right) \\
z_t &= C \cdot y_{t+1} + D \cdot x_{t+1} \quad \left(\text{output}\right)
\end{aligned}
\qquad \text{Mealy}
$$

where A, B, C, and D are matrices, and Y_t, z_t, y_{t+1}, and x_{t+1} are vectors. LSMs have the potential application of describing the interrelationships of a number of simultaneous biochemical reactions, with inputs and outputs feeding into each other. Using the "transfer" function, the gain or attenuation in concentrations of molecular products (and their reaction rates) could be described. Unfortunately, the author in researching the subject could find no reference papers on the use of LSMs in this context. Thus the author regrets that LSMs applied in bioinformatics will not be discussed in this book. Two examples of LSMs are discussed however, in Chapter 11 in Part II of this book[a].

Finite State Automata

A Finite State Automaton (FSA) is a generalization of a sequential machine. The most important theorem about FSA is as follows.

Regular Expressions

0. Regular Expressions (r.e.) may be defined (they are sets of equivalent expressions).

    λ is an r.e.

    ϕ is an r.e.

    if $\alpha \in I$, then α is an r.e.

    if α and β are r.e., then $\alpha \cdot \beta$ is an r.e.

    if α and β are r.e., then $\alpha + \beta$ is an r.e.

    if α is an r.e., then α^* is an r.e.

    Any combination of r.e. using the above is an r.e; else the expression is not an r.e.

---

[a] References for background information may be found in [166, pp. 84-96], and *Linear Sequential Circuits* by A. Gill, McGraw-Hill, 1967.

Deterministic and Non-Deterministic FSAs

Kleene's Theorem:
The set of regular expressions is equivalent to a Deterministic FSA (DFSA).

1. An FSA $M = (S, I, \delta, q_0, F)$, where

S	is a finite number of states;
I	is a finite set of inputs;
$q_0 \in S$	is a unique start state;
$F \subset S$	is a subset of final accepting states;
$\delta: S \times I \to S$	is a relationship that maps the current state and input into the next state.

2. Any Non-Deterministic FSA (NDFSA) has an equivalent DFSA.
    Note: A NDFSA is not the same thing as a probabilistic FSA.

3. Given any finite r.e., an equivalent FSA may be constructed.

4. If a FSA has n states and accepts an r.e. of length n, then the FSA accepts an infinite number of r.e. (pumping theorem for r.e.).

5. Every FSA has a corresponding complete FSA
    (state transitions are defined for every input, for all states).

6. Every FSA has its corresponding complement FSA.

7. FSAs may be decomposed (synthesized) from sub FSAs, as follows.

    Given FSAs A and B, then there is an $A \cdot B$ FSA;

    Given FSAs A and B, then there is an $A + B$ FSA;

    Given FSA A, then there is an A^* FSA;

    Given FSA A, then there is an $\overline{A}$ FSA (complement).

FSAs Constitute a Boolean Algebra

As a consequence of the above, the set of FSAs constitute a Boolean algebra, as with the complement and "+", we have functional completeness. Thus, for example, given A, B are FSAs, then $A \cap B = \overline{\overline{A} + \overline{B}}$, etc.

8. As a consequence of the above, we have: r.e. $\Rightarrow$ FSA
    (given any r.e., there exists a corresponding FSA).

9. McNaughton-Yamada Theorem
    Given any FSA, there exists a corresponding r.e.

10. Combining 8 and 9, r.e. equivalent to a FSA (Kleene's Theorem).

11. For every FSA, there exists its corresponding reduced FSA.

Chomsky Grammars

A grammar is defined as $G = \left(V_N, \; V_T, \; \mathcal{P}, \; S\right)$,

where $V = V_N \cup V_T$, and $V_N \cap V_T = \varnothing$ and

V	is the vocabulary
V_N	is the set of non-terminals
V_T	is the set of terminals
$S \in V_N$,	is the start symbol

$\mathcal{P}$ is the set of production rules: there are four classes of grammars that constitute a hierarchy, and each class has its own characteristic set of production rules.

A language L is a set and may be specified or defined simply as a set. An example is any language defined by a regular expression (which may or may not be finite). However, a language may also be generated by a grammar G, in which case we write L(G), or sometimes L_G.

$$L(G) = \left\{ \omega \; | \; S \overset{*}{\Rightarrow} \omega \;\; \text{and} \;\; \omega \in V_T^* \right\}$$

Chomsky Type 3 Grammars

Type 3 grammars are characterized by production rules of the following two forms:
$$A \Rightarrow a \quad \text{or} \quad A \Rightarrow a\,B \qquad \text{where } a \in V_T, \text{ and } A, B \in V_N.$$
Type 3 grammars are also called regular grammars.

Type 2 grammars are characterized by production rules of the following form:
$$A \Rightarrow \alpha, \text{ where } \alpha \in V^+, A \in V_N, \text{ but } S \Rightarrow \lambda, \text{ is permitted.}$$
Type 2 grammars are also called context-free grammars.

Type 1 grammars are characterized by production rules of the following form:
$$\alpha \, A \, \beta \Rightarrow \zeta, \text{ where } \alpha, \beta \in V^*, A \in V_N, \text{ and } \zeta \in V^*, \text{ and } |\zeta| \geq |\alpha \, A \, \beta|.$$
Type 1 grammars are also called context-sensitive grammars.

Type 0 grammars are characterized by production rules of the following form:
$$\alpha \, A \, \beta \Rightarrow \zeta, \text{ where } \alpha, \beta \in V^*, A \in V_N, \text{ and } \zeta \in V^*.$$
Type 0 grammars are also called unrestricted phrase structured grammars.

There are a number of theorems about context-sensitive languages.

Theorem   If $G = \left(V_N, \; V_T, \; \mathcal{P}, \; S\right)$ is a type 3 grammar,

then there exists a NDFSA $M = \left(K, \; I, \; \delta, \; q_0, \; F\right)$

such that $L(G) = \left\{ \omega \; | \; S \overset{*}{\Rightarrow} \omega \;\; \text{and} \;\; \omega \in V_T^* \right\} = \left\{ \omega \; | \; \delta^*\left(q_0, \; \omega\right) \in F \right\}$

Theorem If $M = \left(K, \ I, \ \delta, \ q_0, \ F\right)$ is a DFSA, then there exists a type 3 grammar $G = \left(V_N, \ V_T, \ \mathcal{P}, \ S\right)$ such that

$$\left\{\omega \ | \ \delta^*\left(q_0, \ \omega\right) \in F\right\} = \left\{\omega \ | \ S \overset{*}{\Rightarrow} \omega \ \text{ and } \ \omega \in V_T^*\right\} = L\left(G\right)$$

With these two theorems, we have the result that the language described by a FSA is equivalent to a Chomsky type 3 grammar. Of course, we also have the Kleene equivalence as well.

Theorem If L is generated by a type 3, type 2, or type 1 grammar,
then $L \cup \{\lambda\}$ and $L - \{\lambda\}$ remains a type 3, type 2, or type 1 language.

Chomsky Type 2 Grammars

Theorem If $G = \left(V_N, \ V_T, \ \mathcal{P}, \ S\right)$ is a type 2 grammar,
then there exists an algorithm to determine if $L(G) = \phi$

Theorem (pumping theorem for context-free languages)

If $G = \left(V_N, \ V_T, \ \mathcal{P}, \ S\right)$ is a type 2 grammar, then there exists a positive integer k such that if $\omega \in L(G)$ and $|\omega| > k$, then there are u, v, w, x, y $\in V_T^*$ where $|vwx| \le k$ and

$|vx| > 0$, such that if $\omega = uvwxy$, then $uv^iwx^iy \in L(G)$ for all $i \ge 0$.

Theorem If $G = \left(V_N, \ V_T, \ \mathcal{P}, \ S\right)$ is a type 2 grammar, then it is possible to construct another type 2 grammar $G^1 = \left(V_N^1, \ V_T^1, \ \mathcal{P}^1, \ S^1\right)$ such that $L\left(G\right) = L\left(G^1\right)$ and

for all $A \in V_N^1$, $A \overset{*}{\Rightarrow} \omega$ and $\omega \in V_T^*$.

Theorem If $G = \left(V_N, \ V_T, \ \mathcal{P}, \ S\right)$ is a type 2 grammar and $L(G) \ne \phi$, then it is possible to construct another type 2 grammar $G^1 = \left(V_N^1, \ V_T^1, \ \mathcal{P}^1, \ S^1\right)$ such that $L\left(G\right) = L\left(G^1\right)$ and for all $A \in V_N^1$, $S^1 \overset{*}{\Rightarrow} \alpha A \beta \overset{*}{\Rightarrow} \alpha \delta \beta$, where $\alpha, \ \beta, \ \delta \in \left(V_T^1\right)^*$.

Theorem If $G = (V_N, V_T, \mathcal{P}, S)$ is a type 2 grammar, then it is possible to construct another type 2 grammar $G^1 = (V_N^1, V_T^1, \mathcal{P}^1, S^1)$ such that $L(G) = L(G^1)$ and all non-terminals V_T^1 generate only an infinity of terminal strings.

Theorem If $G = (V_N, V_T, \mathcal{P}, S)$ is a type 2 grammar, and all production rules are of the form $A \Rightarrow \alpha$, where $\alpha \in V^*$, then it is possible to construct another type 2 grammar $G^1 = (V_N^1, V_T^1, \mathcal{P}^1, S^1)$ such that $L(G) = L(G^1)$ and all production rules in P^1 are of the form $A \Rightarrow \alpha$ where $\alpha \in V^+$ or $S^1 \Rightarrow \lambda$ and S^1 does not appear on the right-hand side of any production rule in $\mathcal{P}^1$.

Theorem If $G = (V_N, V_T, \mathcal{P}, S)$ is a type 2 grammar and $S \underset{G}{\overset{*}{\Rightarrow}} \omega$, then there is a leftmost derivation of ω.

Theorem If $G = (V_N, V_T, \mathcal{P}, S)$ is a type 2 grammar and $A, B \in V_N$, then it is possible to construct another type 2 grammar $G^1 = (V_N^1, V_T^1, \mathcal{P}^1, S^1)$ such that $L(G) = L(G^1)$ and there are no production rules in $\mathcal{P}^1$ of the form $A \Rightarrow B$.

Theorem

If $G = (V_N, V_T, \mathcal{P}, S)$ is a type 2 grammar,

then it is possible to construct another type 2 grammar

$G^1 = (V_N^1, V_T, \mathcal{P}^1, S)$ such that $L(G) = L(G^1)$ and the new grammar

is in Chomsky Normal Form (CNF).

(i.e., all production rules of $\mathcal{P}^1$ are of the form

$A \Rightarrow a$, or $A \Rightarrow BC$, where $A, B, C \in V_N^1$, and $a \in V_T$.

Theorem

If $G = (V_N, V_T, \mathcal{P}, S)$ is a type 2 grammar, then it is possible to construct another

type 2 grammar $G^1 = (V_N^1, V_T, \mathcal{P}^1, S)$ such that $L(G) = L(G^1)$ and the new

grammar is in Greibach Normal form;

i.e., all production rules of $\mathcal{P}^1$ are of the form

$$A \Rightarrow a\,\beta, \text{ where } A \in V_N^1, a \in V_T, \text{ and } \beta \in \left(V_N^1\right)^*.$$

Definition: If $A \Rightarrow \alpha A \beta$ and both $\alpha \neq \lambda$ and $\beta \neq \lambda$, then A is said to be self-embedding.

Definition: If a type 2 grammar has at least one non-terminal which is self-embedding, then the grammar is self-embedding.

Theorem

If $G = (V_N, V_T, \mathcal{P}, S)$ is a type 2 grammar that is non-self-embedding,

then it is possible to construct a type 3 grammar

$$G^1 = (V_N^1, V_T, \mathcal{P}^1, S) \text{ such that } L(G) = L(G^1).$$

Theorem

Context-free languages are closed under union, concatenation, and Kleene star, but not under complementation or intersection. Thus context-free languages do not constitute a Boolean algebra.

Definition: A semi-Dyck Language D_{2n} has type 2 grammar $G_{2n} = \left(\{S\}, \ V_{2n}, \ \mathcal{P}_{2n}, \ S\right)$

where: $V_{2n} = \left\{a_i\right\}_{i=1}^n \ \cup \ \left\{a_i^1\right\}_{i=1}^n$

and $\mathcal{P}_{2n} = \left\{S \Rightarrow SS \mid \lambda\right\} \ \cup \ \left\{S \Rightarrow a_i Sa_i^1\right\}_{i=1}^n$

Homomorphism Defined

Definition (of homomorphism h)

1. $h: \Sigma^* \to \Sigma^*$

2. $h(\lambda) = \lambda$

3. $h(\omega_1 \cdot \omega_2) = h(\omega_1) \cdot h(\omega_2)$

4.   For all $a \in V_T$, $|h(a)| = 1$

5. $h(L) = \left\{ \tau \mid \text{exists a } \omega \in L \text{ and } \tau = h(\omega) \right\}$

Chomsky-Schützenberger Theorem

Theorem (Chomsky-Schützenberger)

Every context-free language L equals a homomorphic image of the homomorphism h of the intersection of a Dyck language D_{2n} and a regular language R, or

$$L = h(D_{2n} \cap R)$$

Turing Machines

Definition: A Turing machine $T = \left(S, \ \Sigma, \ \Gamma, \ \delta, \ q_0, \ F \right)$

       where: S   is a finite number of states

                 $\Sigma \subset \Gamma$ is a finite set of tape input symbols

                 Γ is a finite set of tape output symbols (one is a blank, Δ)

                 $\delta: S \times \Sigma \to S \times (\Gamma - \{\Delta\}) \cup S \times \{ L, R \}$

                 $q_0 \in S$ is a unique start state

                 $F \subset S$ is a subset of final or accepting states

       if δ is single-valued, then T is deterministic; else T is non-deterministic.

       Alternatively, instead of F, $T = \left(S, \ \Sigma, \ \Gamma, \ \delta, \ q_0, \ H \right)$, and

                 $\delta: S \times \Sigma \to S \times (\Gamma - \{\Delta\}) \cup S \times \{ L, R \} \cup H,$

       where: $H \subset S$ is a subset of halting states.

There are indeed, many alternative definitions of Turing machines, including multi-tape, multi-head, multi-dimensional tapes, one-way tapes, and Turing machines as 4-tuples, as 5-tuples, etc. We shall not discuss these, but shall assume that they are known to the reader. We shall also assume that the reader is familiar with the idea of writing a description of a Turing machine on a tape, and running a Universal Turing machine with that description plus an input.

Chomsky Type 0 Languages

Theorem If a language is recursively enumerable (computable by a Turing Machine), then it can be generated by a Chomsky type 0 grammar G.

Theorem If a language L is generated by a Chomsky type 0 grammar G, then there is a Turing machine that accepts L.

Thus Chomsky type 0 languages L are equivalent to Turing machines.

Pushdown Automata

Definition: A Pushdown Automaton $M = \left(S, \ \Sigma, \ \Gamma, \ \delta, \ q_0, \ Z_0, \ F\right)$

where: S   is a finite number of states
Σ is a finite set of tape input symbols
Γ is a finite set of tape (pushdown, stack) output symbols
$\delta: S \times (\Sigma \cup \{\lambda\}) \times \Gamma^* \to S \times \Gamma^*$
$q_0 \in S$ is a unique start state
$F \subset S$ is a subset of final or accepting states
if δ is single-valued, then M is deterministic;
else M is non-deterministic.

As with Turing machines, there is a variety of definitions of Pushdown automata. I assume that the reader is familiar with different definitions, such as:

accepting in a final state

accepting when the input tape is empty,

accepting when in a final state when the input tape is simultaneously empty, etc.

An important note however: just as Turing machines may have multi-dimensional tapes, so may Pushdown automata have multi-dimensional tapes (where the input symbols might be viewed as vectors).

Pushdown Automata and Chomsky Type 2 Languages

Theorem If $G = \left(V_N, \ V_T, \ \mathcal{P}, \ S\right)$ is a type 2 grammar, then it is possible to construct a Pushdown automaton that accepts L(G).

Theorem Given a pushdown automaton $M = \left(S, \ \Sigma, \ \Gamma, \ \delta, \ Q, \ Q, \ \varnothing \right)$, then the language L accepted by M is Chomsky type 2 (there is a $G = \left(V_N, \ V_T, \ \mathcal{P}, \ S \right)$, such that G is type 2.

Thus Chomsky type 2 languages are equivalent to Pushdown automata.

It is known that if one restricts the set of type 2 languages to deterministic type 2 languages, and if the corresponding pushdown automaton is complete (defined for all triples in $\Gamma \times S \times \Sigma$, then the complement Pushdown automaton may be constructed. Thus deterministic type 2 languages are closed with respect to complementation.

Theorem   Pushdown automata $\cap$ FSA = Pushdown automata

Context-Sensitive (Chomsky type 1) Languages

Definition: Production rules for type 1 languages are of the form
$$X_1 X_2 \dots X_n \Rightarrow Y_1 Y_2 \dots Y_m \text{, where } m \geq n,$$
but if all the production rules are of the form in which $0 < n \leq m \leq 2$, then the grammar is in Kuroda Normal Form.

Thus this means that he production rules in KNF look like

$$
\begin{aligned}
X_1 X_2 &\Rightarrow Y_1 Y_2 \\
X_1 &\Rightarrow Y_1 Y_2 \\
X_1 &\Rightarrow Y_1
\end{aligned}
$$

Theorem

A type 1 language in KNF is equivalent to all the production rules being of the form:

$$
\begin{aligned}
A &\Rightarrow a \\
A &\Rightarrow B \\
A &\Rightarrow BC \\
AB &\Rightarrow CD
\end{aligned}
$$

Definition: A B $\Rightarrow$ A C is said to be left sensitive.
A B $\Rightarrow$ C B is said to be right sensitive.

Definition: A type 1 grammar in which all production rules of the form A B $\Rightarrow$ C D are replaced by: A B $\Rightarrow$ A Y, A Y $\Rightarrow$ Z Y, Z Y $\Rightarrow$ Z D, Z D $\Rightarrow$ C D is said to be in refined KNF.

Theorem
Every λ-free type 1 grammar is equivalent to a type 1 grammar in which all production rules are in the following form:

$$
\begin{array}{rcl}
A & \Rightarrow & a \\
A & \Rightarrow & B \\
A & \Rightarrow & BC \\
AB & \Rightarrow & AC \qquad \left(\text{left sensitive}\right) \\
AB & \Rightarrow & BA \qquad \left(\text{permuting}\right)
\end{array}
$$

Linear Bounded Automata

Theorem   Type 1 languages are equivalent to Linear Bounded automata.

(It is assumed that the reader understands that a LBDA is effectively a Turing machine that uses an amount of tape which is a linear function of its input).

Theorem (Ginsburg and Greibach)
If $G = \left(V_N, \; V_T, \; \mathcal{P}, \; S\right)$ is a type 1 grammar, and f is a type 1 substitution on V_T that does not contain λ, then $f\left(L\left(G\right)\right)$ is a type 1 language.

Lindenmeyer (Developmental) $\mathcal{L}$ Systems

Notation: D   deterministic   0L   context-free   1L   context-sensitive

Definition: A D0L system is defined as follows: $G = \left(V_T, \; \mathcal{P}, \; W\right)$

where: V_T is a finite set of (terminal) symbols
$\mathcal{P}$ is a finite set of production rules
W is a finite set of axioms (start expressions)

Note 1: If $\alpha \Rightarrow \beta \; \in \mathcal{P}$ and if there is more than one β, then G is non-deterministic.

Note 2: The replacement (in production rules) takes place in parallel (not serially).

Thus, given D0L system $G = \left(V_T, \; \mathcal{P}, \; W\right)$,

$$
L\left(G\right) = \left\{ h^i\left(\omega\right) \mid i \geq 0 \right\} \text{ where } h \text{ is a homomorphism.}
$$

Shape Grammars

Definition: $SG = (V_M, V_T, \mathcal{R}, I\,)$ is a shape grammar,
V_M is a finite set of non-terminal symbols called markers
V_T is a finite set of terminal shape elements
$\mathcal{R}$: $u \Rightarrow v$ is the finite set of shape rules, $u \in$ mix of $V_T^* \,\& \, V_M^+$, $v \in$ mix of $V_T^* \,\& \, V_M^*$

$I \in V_T^* \times V_M^+ \times V_T^*$ is the initial shape

Note: $V_M \cap V_T = \varnothing$. If the shape rules are applied in parallel (not serially) then this is a parallel shape grammar. See the Appendix for greater detail.

Chapter 3: The Beginning Numbers

Numerology

"In prognosticating the course of disease, learned physicians invoked quasi-mathematical theories about periodicity and favorable or unfavorable days. The original basis for such theories was the frequent inclusion in case descriptions in the Hippocratic *Epidemics* of information about the number of days from the onset of illness to the day on which particular phenomena occurred, as well as the behavior of recurrent fevers such as malaria. Data of this kind, initially derived from observation, were subsequently interpreted in the light of beliefs held in antiquity about the properties of numbers and auspicious or inauspicious calendar dates. Galen's treatises *On Crisis* and *On Critical Days* provided a very full treatment of the whole subject, which was subsequently taken up by various Muslim medical authors and further developed in Latin scholastic medicine.
"Two things determined the nature and the outcome of the crisis: the state of the patient's own body and whether or not the crisis fell on a favorable day. But opinions differed as to the intervals at which favorable days recurred and from what point one should start counting. Thus, the identification of such days and the proper means for determining them became subjects of medical debate.

"Because it involved calendar dates and thus the motions of the moon, the theory of critical days was a branch of medical astrology" [231, p. 135].

"The advocacy of supplicatory prayers with suggested texts and cryptograms is a common feature of many plague treatises for both the prevention and alleviation of plague" [225, p. 123].

"Many of these esoteric practices are related to 'letter magic,' which is the use of the Arabic language against evil ..." [225, p. 122].

"It is advised that a Muslim should repeat the special prayers according to the number of words in the prayer; and if more prayers were needed, the prayer should be repeated according to the number of letters in the prayer" [225, p. 124].

"For example, whoever said: 'The eternal, there is no destruction and cessation of His kingdom' every day 136 times would be saved from the disease. Whoever repeated the various names of God, such as 'the Preserving' every day 898 times or 'the Vigilant' 312 times, would be safe. If a Muslim were devout and repeated 'the Subduer' over the ill 2142 times, plague would depart" [225, p. 128]. It was believed that the plague could be averted by drinking from cups of water with the following sign or talisman written upon it (see the talisman at the left), while the triangular form (at the right) was thought effective during plague epidemics [225, pp. 130, 131].

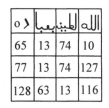

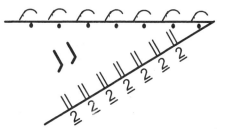

The Beginning: Numbers

It is important, from the point of view of computation, to know and have a clear understanding of the numbers involved in understanding the subject under study. The numbers are as follows: two pairs of Watson-Crick complementary bases: (A/T, C/G for DNA; A/U, C/G for RNA). These four DNA bases support a maximum of $4^3 = 64$ possible codon triplets, which allow (with mRNA and tRNA) the construction of polypeptide chains or proteins, composed of 20 amino acids. These are the numbers we must deal with–or are they? The following discussion, interesting per se, will challenge these numbers. This chapter refers to more biochemistry than most of the rest of the book, but the main ideas should be easy to follow.

Error Detection: Parity

DNA and RNA are complicated molecules. The replication of such molecules, as with any complex process, is subject to error. Errors occur in nature, and of course apply to the replication of DNA and RNA and explain some genetic diseases, as well as innovations that are useful during evolution. The following discussion is quite interesting, in that errors that may take place can be detected using a simple parity check [11], [99], [171].

It is first necessary to examine not only the four nucleotide bases, but other nucleotide bases that might also have been possible candidates in the genetic code. These bases will be viewed from the point of view of hydrogen bonding: donors and acceptors. In addition, we must consider steric "stacking," the pairing of purine with pyrimidine.

Using the classification by Mac Dónaill, an acceptor will be signified by "0", and a donor will be signified by a "1". In addition, a purine derivative will be signified by "R" (from puRine) where R is equivalent to "0", and a pyrimidine derivative will be signified by "Y" (from pYrimidine) where Y is equivalent to "1". This information may be encoded in binary as four bits, the first three bits representing the donor/acceptor bonding sites of a nucleotide and the last bit encodes for purine or pyrimidine. We will find, for example:

$$C = \langle 100,\ 1 \rangle \text{ while } G = \langle 011,\ 0 \rangle$$

These are complements, or $\quad \overline{C \oplus G} = \overline{\langle 111,\ 1 \rangle} = \langle 000,\ 0 \rangle$, and the Hamming

distance is $\partial\left(\overline{C \oplus G}\right) = \partial\left(\overline{\langle 111,\ 1 \rangle}\right) = \partial\left(\langle 000,\ 0 \rangle\right) = 0$. Note that Mac Dónaill

uses $C \oplus G$, not $\overline{C \oplus G}$, encoding the idea of Watson-Crick complements.

We find the following.

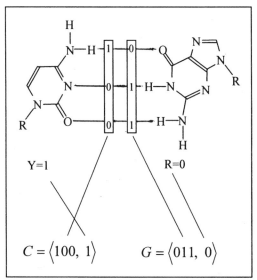

Figure 3.1 Parity: C and G pairs

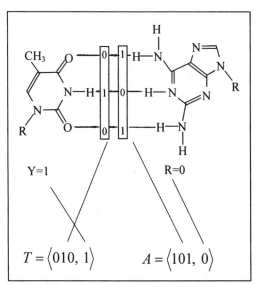

Figure 3.2 Parity: T and A pairs

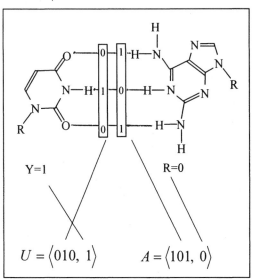

Figure 3.3 Parity: U and A pairs

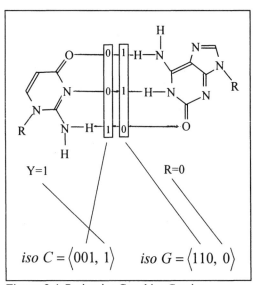

Figure 3.4 Parity: iso C and iso G pairs

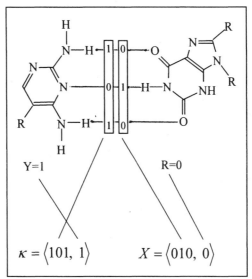

Figure 3.5 Parity: κ and X pairs

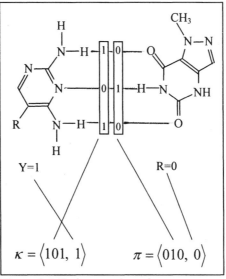

Figure 3.6 Parity: κ and π pairs

The base π is discussed in [11, pp. 33-37].

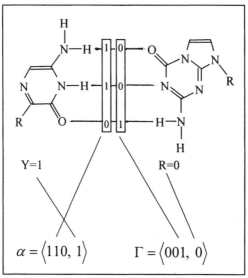

Figure 3.7 Parity: α and Γ pairs

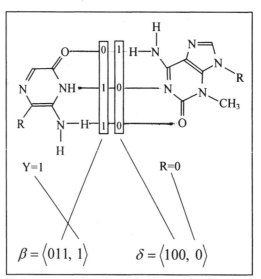

Figure 3.8 Parity: β and δ pairs

If we collect the above information, we find that the rightmost bit acts as a parity check and thus the following.

$$
\begin{array}{l}
\text{naturally} \\
\text{occurring}
\end{array}
\left\{
\begin{array}{ll}
C = \langle 100,\ 1 \rangle & G = \langle 011,\ 0 \rangle \\
\\
U\,/\,T = \langle 010,\ 1 \rangle & A = \langle 101,\ 0 \rangle
\end{array}
\right.
$$

$$
\begin{array}{ll}
iso\ C = \langle 001,\ 1 \rangle & iso\ G = \langle 110,\ 0 \rangle
\end{array}
\right\}\ \text{even parity}
$$

$$
\begin{array}{ll}
\kappa = \langle 101,\ 1 \rangle & X\,/\,\pi = \langle 010,\ 0 \rangle \\
\\
\alpha = \langle 110,\ 1 \rangle & \Gamma = \langle 001,\ 0 \rangle \\
\\
\beta = \langle 011,\ 1 \rangle & \delta = \langle 100,\ 0 \rangle
\end{array}
\right\}\ \text{odd parity}
$$

Figure 3.9

Experiments with Alien Forms of Life

We find that a parity check exists in naturally occurring nucleotide bases in DNA and RNA that acts as an error-detecting mechanism. It has been pointed out [171] that mismatch repair during template replication is also a possibility, thus not only has error detection evolved, but even the possibility of the evolution of error correction. Experiments incorporating new nucleotide bases into DNA or RNA concern new forms, or alien forms of life, and also throw light upon the evolution of biochemistry and genetics, as well as upon biology.

⊕	C	G	A	T/U	iso C	iso G	κ	X/π	α	Γ	β	δ
C	—	—	—	—	—	—	—	—	—	—	—	—
G	1111	—	—	—	—	—	—	—	—	—	—	—
A	0011	1100	—	—	—	—	—	—	—	—	—	—
T/U	1100	0011	1111	—	—	—	—	—	—	—	—	—
iso C	1010	0101	1001	0110	—	—	—	—	—	—	—	—
iso G	0101	1010	0110	1001	1111	—	—	—	—	—	—	—
κ	0010	1101	0001	1110	1000	0111	—	—	—	—	—	—
X/π	1101	0010	1110	0001	0111	1000	1111	—	—	—	—	—
α	0100	1011	0111	1000	1110	0001	0110	1001	—	—	—	—
Γ	1011	0100	1000	0111	0001	1110	1001	0110	1111	—	—	—
β	1110	0001	1101	0010	0100	1011	1100	0011	1010	0101	—	—
δ	0001	1110	0010	1101	1011	0100	0011	1100	0101	1010	1111	—

Figure 3.10 Collected parity information for a number of bases

Hamming distance is given as: $\partial(M \oplus N)$ or the number of ones in $M \oplus N$, and as $M \oplus N$, and as $M \oplus M = N \oplus M$, and as $M \oplus M = 0$, only 66 entries of the 144-entry table are significant. For example, pairs with Hamming distance 1, 2, etc. are easily determined. To pick out Watson–Crick complements, $M \overline{\oplus} N$ is easier, using a Hamming distance measured by $\partial(M \overline{\oplus} N) = 4 - \partial(M \oplus N)$. Thus for a Hamming distance of 0, we get the following complements.

$$
\begin{array}{c|c|c}
G,\ C & iso\ G,\ iso\ C & \Gamma,\ \alpha \\
A,\ T/U & \kappa,\ X/\pi & \delta,\ \beta
\end{array}
$$

$\uparrow$

naturally occurring Watson-Crick complements

Using a Hamming distance computed as $\partial\!\left(M\overline{\oplus}N\right)=4-\partial\!\left(M\oplus N\right)$ to find all the pairs of distance 1, we obtain the following.

$$
\begin{array}{ll}
\kappa,\ G & X/\pi,\ C \\
\kappa,\ iso\ G & X/\pi,\ iso\ C \\
\kappa,\ T/U & X/\pi,\ A \\
\hline
\alpha,\ G & \beta,\ C \\
\alpha,\ iso\ C & \beta,\ iso\ G \\
\alpha,\ A & \beta,\ A \\
\hline
\delta,\ G & \Gamma,\ C \\
\delta,\ iso\ C & \Gamma,\ iso\ G \\
\delta,\ T/U & \Gamma,\ T/U \\
\delta,\ \kappa &
\end{array}
$$

Considering the donor/acceptor triplets, we have all eight possible triplets except $\langle 000\rangle$ and $\langle 111\rangle$, and $\langle 000\rangle$ corresponds to a nucleotide that is unstable due to easy hydrolysis. Other possible modified forms of nucleotide derivatives are considered [171] to determine, by studies of energy and kinetics, the likelihood of error detection and repair (error correction), including:

A^1 A derivative with no amino group at the 2 position
G^1 G derivative with no amino group at the 2 position
δ^1 δ derivative with no amino group at the 2 position
β^1 β derivative with no amino group

Assuming an error is detected in one strand of DNA, and assuming a low probability of an additional error at the same location in the second strand of complements, the second strand may then be used to correct detected errors. In fact, if sequences of bases have been excised, the complementary strand may be used to recreate the excised portion of DNA [89, p. 609] . Aside from error detection and repair, these Watson-Crick base pairs have other important implications, that will now be discussed.

Alien Nucleotide Bases Bound as Chelates

The standard Watson-Crick DNA bases are A/T, and C/G, but in addition to these standard base pairs, self-complementary bases also exist: Metallo-DNA. Artificial nucleotide bases X act like Watson-Crick complementary bases, except X is a self-complement. The hydrogen bonds between nucleotide bases are replaced by a metal-containing chelate coordinated with X (DNA usually acts as a chelating agent) [174].

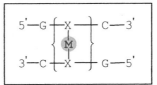

Figure 3.11 Chelated self-complementing bases

Example 1: X in Figure 3.11 is H (abbreviates hydroxypyridone, not hydrogen), and M is copper [177].

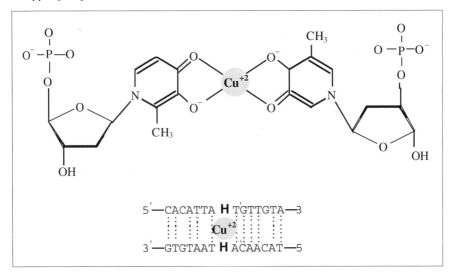

Figure 3.12 Copper chelate of self-complementing hydroxypyridone

From one to five adjacent hydroxypyridone nucleobases have incorporated in a DNA duplex, using copper for added stability.

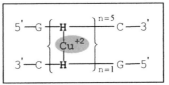

Figure 3.13 Multiple adjacent self-complement bases

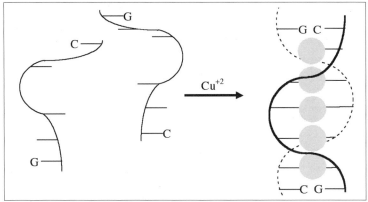

Figure 3.14 Multiple adjacent self-complement bases

Example 2: X is E (abbreviates phenylenediamine [174]),
   M is Ag^+, Cu^{+2}, Pd^{+2}, Pt^{+2}, Ni^{+2}, or
   X is P (abbreviates pyridine [175]), M is Ag^+, Cu^{+2}, Ni^{+2}, Pd^{+2}, or Hg^{+2}, or
   X is B (abbreviates 2, 2'-bipridine), using Cu^{+2} [183].
   Note that X refers to Figure 3.11 in all cases in this example.

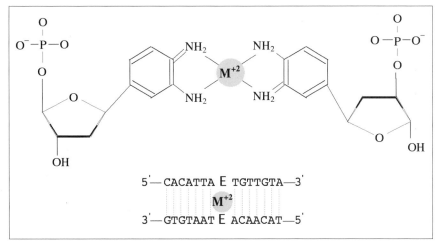

Figure 3.15 Various chelates of self-complementing phenylenediamine

The method of using chelation may also be extended to more complex molecules that are not self-complementing [89]. Some of these modified DNAs may be useful as anti-cancer strategies. More complicated molecules have fluorescent properties.

Example 3: 2, 6-dipicolinate Dipic and pyrimidine Py, using copper coordination [149], [151].

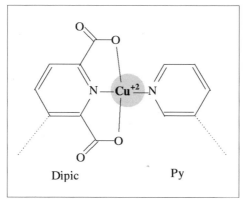

Figure 3.16 Copper chelated base pair

Nucleotide Bases Viewed as Complex Numbers

Other interesting approaches that have implications with regard to parity have been proposed [100], [101]. We can make the following mapping into the complex numbers.

$$
\begin{aligned}
A &\rightarrow i \\
C &\rightarrow -1 \\
G &\rightarrow 1 \\
T/U &\rightarrow -i
\end{aligned}
$$

Note: A/T   maps to a sum of $i + -i = 0$,

         A/U   maps to a sum of $i + -i = 0$,

         C/G   maps to a sum of $1 + -1 = 0$.

Thus the idea of purine/pyrimidine matching is reflected in a zero sum (even parity). However, there is more!

We can do a special mapping, namely, the mapping created when multiplying by -1.

$$
\begin{array}{ccccc}
X & & \rightarrow & -X & \\
A & i & \rightarrow & -i & = & T/U \\
C & -1 & \rightarrow & 1 & = & G \\
G & 1 & \rightarrow & -1 & = & C \\
T/U & -i & \rightarrow & i & = & A
\end{array}
$$

This maps each base to its matching base. Other mappings are also interesting, such as conjugacy, multiplying by i, etc. We can then form vectors associated with a string of bases (5' to 3' direction). For example:

$5'-$G A G T C C G A$-3'$ is transcribed into the following mRNA:

DNA   mRNA

$$
\begin{pmatrix} G \\ A \\ G \\ T \\ C \\ C \\ G \\ A \end{pmatrix} \longrightarrow \begin{pmatrix} C \\ U \\ C \\ A \\ G \\ G \\ C \\ U \end{pmatrix} = \begin{pmatrix} -1 \\ -i \\ -1 \\ i \\ 1 \\ 1 \\ -1 \\ -i \end{pmatrix}
$$

Figure 3.17 Vector for parity representations using complex numbers

Matrices may be constructed to show RNA or DNA structural changes. We shall consider these topics in a later chapter, however.

Aminoacyl-tRNA Synthetases and Alien Nucleotide Bases

aminoacyl - tRNA synthetases

Aminoacy - tRNA Synthetase	*Abbreviation*

Examples:

	tyrosyl - tRNA synthetase	TyrRS
	valyl - tRNA synthetase	ValRS
	isoleucyl - tRNA synthetase	IleRS
	cysteyl - tRNA synthetase	CysRS

amino acid + aminoacyl - tRNA synthetase $\rightarrow$ tRNA$^{amino\ acid}$

Note: tRNA$^{amino\ acid}$ is tRNA capable of (usually charged with) carrying the amino acid, to be emplaced in a protein (in an increasing polypeptide chain). To be more explicit:

amino acid type + aminoacyl - tRNA synthetase $\rightarrow$ amino acid type – tRNA$^{amino\ acid}$

where: amino acid type – tRNA$^{amino\ acid}$ means that "amino acid type" is bonded to tRNA

Examples:

tyrosine + TyrRS + tRNA $\rightarrow$ tRNA^{Tyr} (or tyrosine – tRNA^{Tyr})
valine + ValRS + tRNA $\rightarrow$ tRNA^{Val} (or valine – tRNA^{Val})

A degree of precision may be added with the following view of an aminoacyl - tRNA. [a]

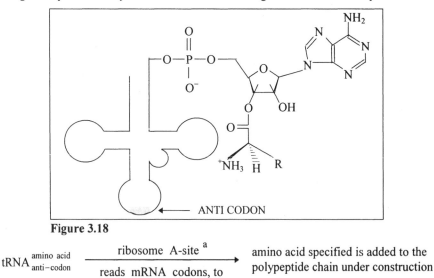

Figure 3.18

$$tRNA^{amino\ acid}_{anti-codon} \xrightarrow[\substack{\text{reads mRNA codons, to} \\ \text{determine the amino acid}}]{\text{ribosome A-site [a]}} \substack{\text{amino acid specified is added to the} \\ \text{polypeptide chain under construction}}$$

---

[a] There are at least two ribosomal binding sites: the A-site accepts the aminoacyl-tRNA carrying the next amino acid residue to be incorporated into the growing peptide chain, and the P-site, which accommodates the tRNA acylated with the growing peptide chain itself [29].

Amino Acids Incorported into Proteins

The important point is to note that the active site of several of the aminoacyl-tRNA synthetases lack the capacity to discriminate between closely similar amino acids. Thus amino acids that do not normally appear in proteins can be expressed in proteins. In addition, these amino acids that do not normally appear in proteins have been shown to exist in living bacteria, and these unnatural proteins contain significant amounts of these abnormal amino acids. Thus from the point of view of computation, the possibility of more than 20 amino acids being incorporated in proteins is quite real. The methodology is to suppress the normal polymerization-terminating activity of the nonsense stop codons, and instead to associate the nonsense codon with a tRNA charged with a "new" amino acid. What are some of these candidate amino acids that do not naturally appear in proteins of life?

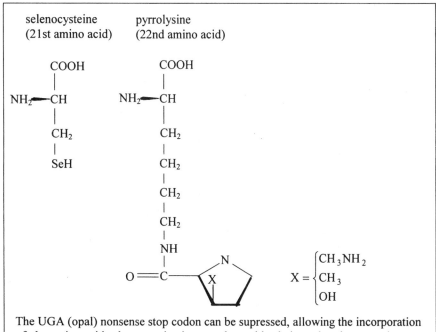

The UGA (opal) nonsense stop codon can be supressed, allowing the incorporation of the amino acid selenocysteine into a polypeptide chain, rather than stopping polymerization [140].

The UAG (amber) nonsense stop codon can be supressed, allowing the incorporation of the amino acid pyrrolysine into a polypeptide chain [6].

These amino acids correspond to stop codons in nature, even if this is not common. Two of the three stop codons may be used in this manner [13].

Figure 3.19

Non-natural amino acid analogs that have been added
to the list of amino acids incorporated into polypeptides

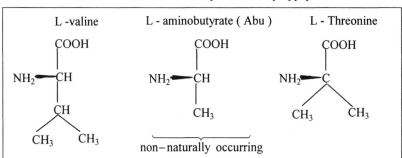

Figure 3.20

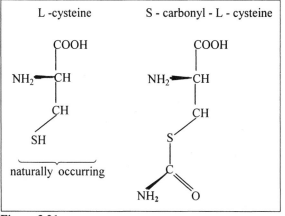

Figure 3.21

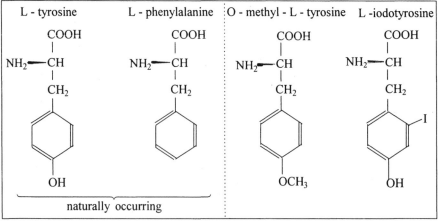

Figure 3.22

In the two very clear examples that follow, nonsense stop codons use mutants to suppress these stop codons, and instead, to insert the amino acid charged by tRNA into the growing polypeptide chain.

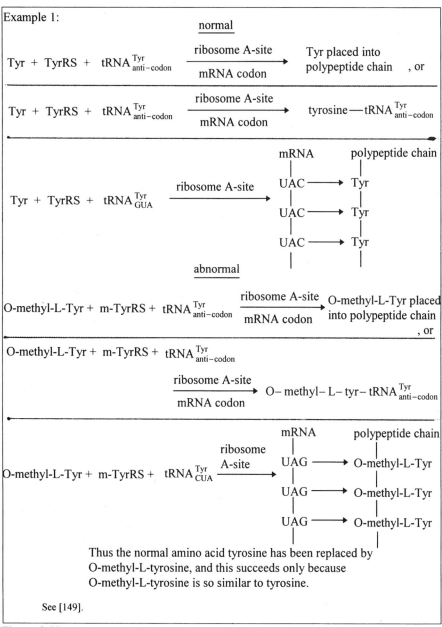

Example 1:

$$\text{Tyr} + \text{TyrRS} + \text{tRNA}^{\text{Tyr}}_{\text{anti-codon}} \xrightarrow[\text{mRNA codon}]{\text{ribosome A-site}} \begin{array}{l}\text{Tyr placed into} \\ \text{polypeptide chain}\end{array} \text{, or}$$

$$\text{Tyr} + \text{TyrRS} + \text{tRNA}^{\text{Tyr}}_{\text{anti-codon}} \xrightarrow[\text{mRNA codon}]{\text{ribosome A-site}} \text{tyrosine} - \text{tRNA}^{\text{Tyr}}_{\text{anti-codon}}$$

$$\text{Tyr} + \text{TyrRS} + \text{tRNA}^{\text{Tyr}}_{\text{GUA}} \xrightarrow{\text{ribosome A-site}}$$

mRNA	polypeptide chain
UAC →	Tyr
UAC →	Tyr
UAC →	Tyr

abnormal

$$\text{O-methyl-L-Tyr} + \text{m-TyrRS} + \text{tRNA}^{\text{Tyr}}_{\text{anti-codon}} \xrightarrow[\text{mRNA codon}]{\text{ribosome A-site}} \begin{array}{l}\text{O-methyl-L-Tyr placed} \\ \text{into polypeptide chain}\end{array} \text{, or}$$

$$\text{O-methyl-L-Tyr} + \text{m-TyrRS} + \text{tRNA}^{\text{Tyr}}_{\text{anti-codon}}$$
$$\xrightarrow[\text{mRNA codon}]{\text{ribosome A-site}} \text{O-methyl-L-tyr} - \text{tRNA}^{\text{Tyr}}_{\text{anti-codon}}$$

$$\text{O-methyl-L-Tyr} + \text{m-TyrRS} + \text{tRNA}^{\text{Tyr}}_{\text{CUA}} \xrightarrow[\text{A-site}]{\text{ribosome}}$$

mRNA	polypeptide chain
UAG →	O-methyl-L-Tyr
UAG →	O-methyl-L-Tyr
UAG →	O-methyl-L-Tyr

Thus the normal amino acid tyrosine has been replaced by O-methyl-L-tyrosine, and this succeeds only because O-methyl-L-tyrosine is so similar to tyrosine.

See [149].

Figure 3.23

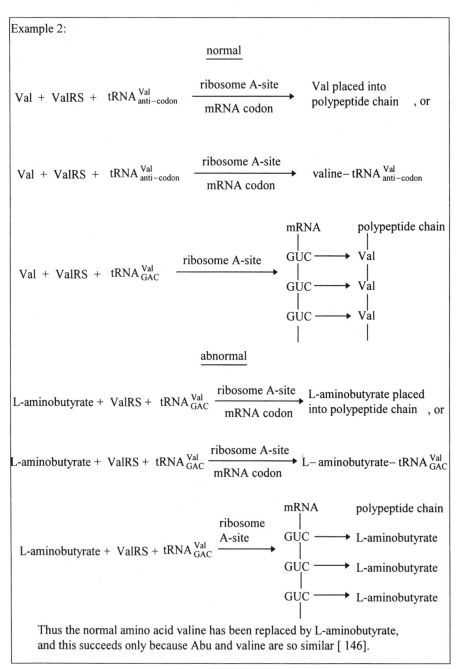

Example 2:

normal

$$\text{Val} + \text{ValRS} + \text{tRNA}_{\text{anti-codon}}^{\text{Val}} \xrightarrow[\text{mRNA codon}]{\text{ribosome A-site}} \text{Val placed into polypeptide chain} \quad , \text{ or}$$

$$\text{Val} + \text{ValRS} + \text{tRNA}_{\text{anti-codon}}^{\text{Val}} \xrightarrow[\text{mRNA codon}]{\text{ribosome A-site}} \text{valine} - \text{tRNA}_{\text{anti-codon}}^{\text{Val}}$$

$$\text{Val} + \text{ValRS} + \text{tRNA}_{\text{GAC}}^{\text{Val}} \xrightarrow{\text{ribosome A-site}}$$

mRNA	polypeptide chain
GUC $\longrightarrow$	Val
GUC $\longrightarrow$	Val
GUC $\longrightarrow$	Val

abnormal

$$\text{L-aminobutyrate} + \text{ValRS} + \text{tRNA}_{\text{GAC}}^{\text{Val}} \xrightarrow[\text{mRNA codon}]{\text{ribosome A-site}} \text{L-aminobutyrate placed into polypeptide chain} \quad , \text{ or}$$

$$\text{L-aminobutyrate} + \text{ValRS} + \text{tRNA}_{\text{GAC}}^{\text{Val}} \xrightarrow[\text{mRNA codon}]{\text{ribosome A-site}} \text{L-aminobutyrate} - \text{tRNA}_{\text{GAC}}^{\text{Val}}$$

$$\text{L-aminobutyrate} + \text{ValRS} + \text{tRNA}_{\text{GAC}}^{\text{Val}} \xrightarrow{\text{ribosome A-site}}$$

mRNA	polypeptide chain
GUC $\longrightarrow$	L-aminobutyrate
GUC $\longrightarrow$	L-aminobutyrate
GUC $\longrightarrow$	L-aminobutyrate

Thus the normal amino acid valine has been replaced by L-aminobutyrate, and this succeeds only because Abu and valine are so similar [146].

Figure 3.24

More than 64 Codons are Possible

Using parity provided by purine/pyrimidine Watson-Crick complementary pairs, along with hydrogen donor/acceptor patterns, then at least six base pairs are possible. Add to this chelated self-complements, as well as chelated pairs, abasic pairs, F, Z (see Figure 3.31), N and H (see Figure 3.32), then many more bases than 4, as well as many more codons, are possible than the standard view of the maximum $4^3 = 64$ usually considered. Thus being limited to just the 16 nucleotide bases of the following 17 A/U, C/G, iso C/iso G, κ/χ or κ/π, α/Γ, β/δ, Hydroxypyridone, phenylEnediamine, Dipic/Pyrimidine, and assuming that any three bases can appear in justaposition, then already $16^3 = 4096$ triplet codons are possible (with redundancies of course). In fact [13], iso-C, A, G is the 65th codon (with iso-G, U, C its anti-codon). This 65th codon/anti-codon pair was chosen to incorporate idotyrosine as its corresponding non-natural amino acid.

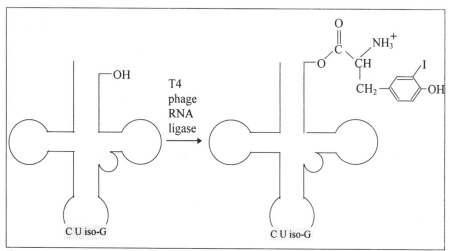

Figure 3.25

Note: DNA polymerase I is very useful. DNA polymerase I may be cleaved using a restriction enzyme into two components, the larger of which is referred to as the Klenow fragment. The Klenow fragment retains polymerase activity, and has been used to create non-standard nucleobases and variants of standard nucleobases (for example, to incorporate iso-G [14, p. 1313]. Other polymerases are used as well, such as the T7 DNA polymerase, used to incorporate iso-C) [12].

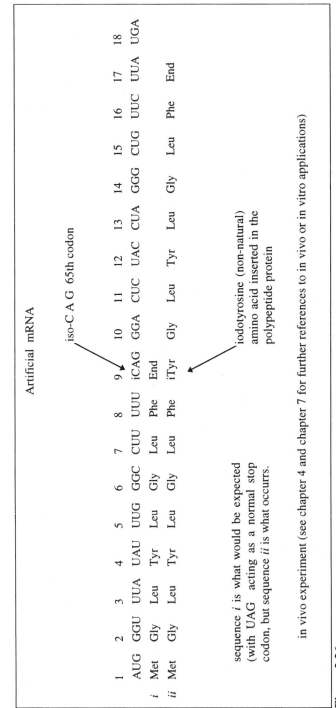

Figure 3.26

Extended Watson-Crick Complements

Referring to Figure 3.26, it has been pointed out [13, p. 161]: "This was the first time that an enzyme process had been observed where the non-standard base pair was accepted as readily as standard bases" and "These observations suggest that an expanded genetic lexicon could be implemented in an in vivo cell culture system ..."

Note: we can thus imagine a DNA double helix with an expanded lexicon that might appear as follows.

Given Watson-Crick complementary base
pairs for an expanded lexicon such as:
A/T, C/G, M/P, Q/R, X/Y

DNA

... A C T G A X G M R T Y ...
... T G A C T Y C P Q A X ...

the corresponding mRNA being:

... U G A C U Y C P Q A X ...

Figure 3.27

To a degree, this has been accomplished [14].

Researchers have found modified base pairs that still fulfill Watson-Crick complementarity and that have the ability to function in DNA. However, is it possible to generalize on this feature, and to construct base pairs that satisfy Watson-Crick complementation, yet are entirely artificial or new [150]? Consider the following.

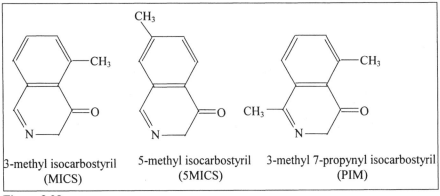

3-methyl isocarbostyril
(MICS)

5-methyl isocarbostyril
(5MICS)

3-methyl 7-propynyl isocarbostyril
(PIM)

Figure 3.28

Then we find that MICS and 5MICS do work, at least in a limited way, as Watson-Crick complementary base pairs, as follows.

	N	X
5^1 GCGTACXCATGCG	A	MICS
 ⋮	T	MICS
3^1 CGCATGNGTACGC	C	MICS
	G	MICS
	A	5MICS
5^1 GCGTACNCATGCG	T	5MICS
 ⋮	C	5MICS
3^1 CGCATGXGTACGC	G	5MICS

Figure 3.29

MICS and 5MICS work with DNA polymerase substrates. Additional work [87] indicates that possible bases (with varying complements, and with mismatched base complements reported too) are provided by such structures still supporting the double-helix structure with base analogues that are non-polar (hydrogen-bonding not a factor, but retaining stacking). See Figures 3.30 through 3.32.

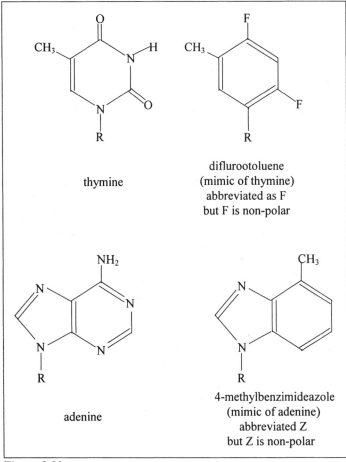

Figure 3.30

It has been pointed out [88] that the non-polar base mimics F and Z (being non-polar, thus hydrogen bonding in the double helix not being a factor); yet A/F and F/A as well as T/Z pairs are processed at almost the same frequency as A/T and T/A base pairings (although there is an asymmetry in the synthesis rates of A/F and F/A).

Thus the polymerase synthesis (creation) of double-helices such as:

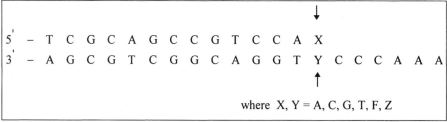

Figure 3.31

A number of other non-natural mimics are mentioned [86]. Two other such molecules include "N" (not symbolizing Nitrogen) and "H" (not symbolizing Hydrogen), as follows:

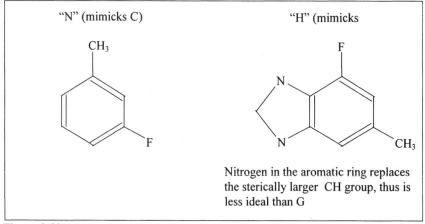

Figure 3.32

Thus purine and pyrimidine shapes (steric stacking effects) may not be as significant as is thought. This suggested further research [84].

Considering a triphosphate deoxy pyrene nucleoside matched against an abasic ribose, noting that pyrene β - deoxynuclease approximates the size of a pyridine/purine pair:

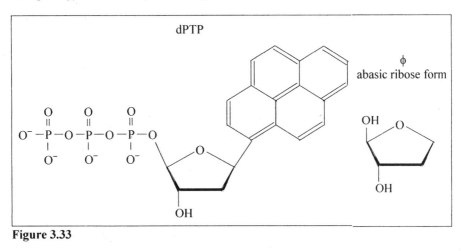

Figure 3.33

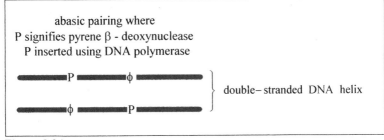

Figure 3.34

Thus in addition to the standard DNA Watson-Crick base pairs A/T, C/G, we must also consider:

mismatched base pairs such as	C/A, C/T,
non-polar non-natural base pairs such as	A/F, T/Z, iso-C/iso-G
self-complementary bases, other chelates	H/H
abasic non-natural bases such as	P/φ
triple, quadruple, and quintuple strands of DNA	

In most cases, we shall consider only standard double helices, composed only of the standard naturally occurring nucleoside bases A, C, G, and T, without mismatches, but we should bear in mind that at least from the computational point of view of chemistry and biochemistry, other possibilities must be considered when dealing with DNA, RNA, and polypeptide proteins.

Alternative Nucleic Acid Polymers

It is possible that the earliest organisms on Earth (and possibly life outside the Earth) might be organized around molecules other than DNA and RNA. To be explicit, rather than using ribose or deoxyribose, threose has some interesting properties.

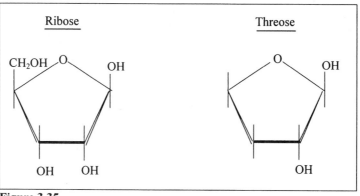

Figure 3.35

What are some of the properties that threose has that are interesting?

1. TNA (L - α - threosefuranosyl oligonucleotides), polymers with phospho-diester bonds connected to nucleotides ($3^1 \rightarrow 2^1$) exist. TNA is analogous to DNA.

2. TNA forms stable Watson-Crick double helices.

3. TNA has Watson-Crick complementary base pairing.

Thus these polymers look like the following (see Figure 3.36).

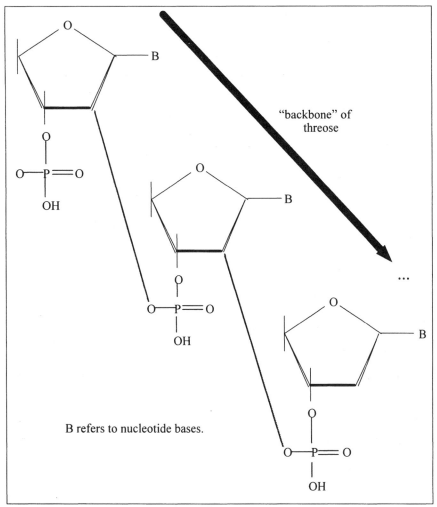

"backbone" of
threose

B refers to nucleotide bases.

Figure 3.36

In addition to threose, some hexoses also form polymers with nucleotides, and also form double helices. We shall not consider this topic further. The conclusion of the discussion in this chapter is that while the basic numbers used include four DNA bases (A, C, G, T, with U replacing T in RNA), 64 possible codons, and 20 amino acids, all these numbers are somewhat provisional, and even the molecules may not be limited to DNA, RNA, and amino acid peptides! Even if life forms are so limited, chemical possibilities need not be so limited, and are thus of interest [117], [51].

To summarize, we have more possible nucleotide bases than the usual four. We already have seen A/U, C/G, iso C/iso G, κ/χ or κ/π, α/Γ, β/δ, Hydroxypyridone, phenylEnediamine, Dipic/Pyrimidine, abasic pairs, MICS, 5MICS (see Figure 3.28), F, Z (see Figure 3.31), and N and H (see Figure 3.32). We thus have more than the usual $4^3 = 64$ base triple codons (even a 65th functioning codon), rather thousands of new possible codons. We have more than the standard 20 amino acids, already finding selenocysteine (21st), pyrrolysine (22nd), L-aminobutyrate, S-carbonyl-L-cysteine, O-methyl-L-tyrosine, L-iodotyrosine, etc.

This chapter is about numbers: the number of nucleotide bases, the number of codons, the number of amino acids. Whatever numbers are used, be aware that any serious treatment of the biochemistry of DNA, RNA, and proteins (at least in the laboratory) may require a broader outlook than the numbers most often assumed. We know clearly that the number of nucleotide bases in DNA and RNA may exceed 4, that there may be more than 64 codons, and that more than 20 amino acids may be constituents of proteins. With this knowledge, we may start to examine aspects of the emergent theory of computation found in bioinformatics. Any theory of linguistic computation must be able to deal with an increasing number of possible bases, codons, and proteins with an increasing number of constituent amino acids.

The first example of the reality of computability in bioinformatics that startled many people was Adleman's in vitro experiment. While details about this experiment can be examined in detail, we shall not do so here, as this experiment is in many respects not representative of the typical problems we shall encounter in this book. However, the experiment focuses upon solving the Hamiltonian path problem, a problem from graph theory. The problem may be reviewed in almost any book dealing with graph theory, but the point is that this problem is well known to be NP-complete (unavoidably much computer time must be expended to solve problems in this class, and all the problems in this class have been shown to be computationally equivalent). 20-mer nucleotide tag sequences are used in each vertex to provide "from" and "to" addresses. Thus two 20-mers in order, provide a route through the Hamilton graph. Allowing polymerases and ligations, the set of all paths traversed can be derived, and specifically, DNA molecules were isolated that corresponded to Hamilton paths. Thus Adleman used strands of DNA in a "test tube" environment to obtain, in parallel fashion, solutions to this problem: an aspect of computation was implemented using DNA. In the next few chapters, we shall make a detailed examination of different aspects of bioinformatics, about which I hope the reader will enjoy learning.

Chapter 4: Regular Languages: DNA and RNA

Epizootics[a]

"In 1317 and 1318, harvests throughout Europe improved, and conditions gradually got better. But a new catastrophe, animal murrains, began. From 1316 to 1322, a series of livestock epidemics devastated what remained of Europe's cattle population. The next two years, 1322 and 1323, proved to be a period of respite, but they were followed by a succession of sheep murrains in 1324 and 1325" [227, p. 29].

The term epizootic refers to an event when a disease attacks a significant number of animals simultaneously or is prevalent among a group of animals within a specified geographic range.

"It is a remarkable story that I have to relate. And were it not for the fact that I am one of many people who saw it with their own eyes, I would scarcely dare believe it, let alone commit it to paper, even though I had heard it from a person whose word I could trust. The plague I have been describing was of so contagious a nature that very often it visibly did more than simply pass from one person to another. In other words, whenever an animal other than a human being touched anything belonging to a person who had been stricken or exterminated by the disease, it not only caught the sickness, but died from it almost at once. To all of this, as I have just said, my own eyes bore witness on more than one occasion. One day, for instance, the rags of a pauper who had died from the disease were thrown into the street, where they attracted the attention of two pigs. In their wonted fashion, the pigs first of all gave the rags a thorough mauling with their snouts after which they took them between their teeth and shook them against their cheeks. And within a short time they began to writhe as though they had been poisoned, then they both dropped dead to the ground, spreadeagled upon the rags that had brought about their undoing" [228, p. 28].

---

[a] Epidemic, demos; epizootic zoology

FSA and Regular Languages: DNA and RNA

DNA and RNA may be studied from the viewpoint of FSA and regular languages. This chapter will attempt an analysis of DNA and RNA from the viewpoint of FSA and regular or Chomsky type 3 languages.

Cut Grammars

An approach to dealing linguistically with DNA and RNA called "cut grammars" [164], will now be discussed.

RNA are single-stranded molecules that fold to simulate double strands like those found in DNA, thereby gaining stability through base-pair and hydrogen bonding. Hence, RNA uses intermolecular bonding to gain stability, while DNA gains stability through the same kind of bonding, but between two strands of molecules, thus through intermolecular forces.

A new type of grammar is proposed that models intermolecular structures, called a "cut grammar."

Definition of a cut grammar:

$$G = \left(V_N, \ V_T, \ \mathcal{P}, \ S\right) \text{ where the production rules } \mathcal{P} \text{ are defined as follows.}$$

$$\left(V_N \cup V_T\right)^* V_N \left(V_N \cup V_T\right)^* \times \left(V_N \cup V_T \cup \{\delta\}\right)^* \quad \text{and} \quad \delta \notin V_N \cup V_T$$

Note: As $\delta \notin \left(V_N \cup V_T\right)^* V_N \left(V_N \cup V_T\right)^* \times \left(V_N \cup V_T \cup \{\delta\}\right)^*$, then δ does not appear in any context (left-hand side of any production rule)

Definition of a cut language:

$$L(G) = \left\{\omega \ | \ \omega \in \left[V_T \cup \{\delta\}\right]^* \ \& \left(S \overset{*}{\Rightarrow} \omega\right)\right\}$$

Definition: Given ω is a string in a cut language and $\omega = \omega_1 \delta \omega_2 \delta \ \ldots \ \omega_n \delta$ and $\omega_i \in V_T^*$ then

1. $\hat{\omega} = \left\langle \omega_1, \omega_2, \ldots, \omega_n \right\rangle$ an ordered tuple of strings with all instances of δ cut out. thus $\hat{\omega}$ is ω cut into pieces at all instances of δ, order preserved.

2. $\tilde{\omega} = \omega_1 \omega_2 \ldots \omega_n$ is ω with all instances of δ removed, a single string.

Definition:

$$\hat{L}(G) = \left\{ \hat{\omega} \mid \hat{\omega} \in 2^{V_T^*} \ \& \ S \overset{*}{\Rightarrow} \omega \right\}$$

$$\tilde{L}(G) = \left\{ \tilde{\omega} \mid \tilde{\omega} \in V_T^* \ \& \ S \overset{*}{\Rightarrow} \omega \right\}$$

Note that the definitions given can be modified to support all Chomsky type languages, specifically regular, context-free, and context-sensitive.

Restriction Languages

Recall that restriction enzymes cut DNA at restriction sites; thus MboI cuts as follows.

$$
\begin{array}{cccc}
\downarrow & & & \\
\underline{g} & a & t & c \\
\vdots & \vdots & \vdots & \vdots \\
\underline{c} & t & a & g \\
& & & \uparrow
\end{array}
$$

Regular (Chomsky Type 3) Cut Grammars

$$G_p = \left(\{S\}, \ \{a, \ c, \ g, \ t\}, \ \mathcal{P}_p, \ S \right) \quad \text{(the "p" subscript refers to a partial digest)}$$

where: $\mathcal{P}_p = \left\{ S \ \Rightarrow \ aS \mid cS \mid gS \mid tS \mid \delta \, gatcS \mid \lambda \right\}$

Two sample derivations:

$$
\begin{aligned}
S \ &\Rightarrow \ aS \Rightarrow atS \Rightarrow atgS \Rightarrow atgcS \Rightarrow atgc \, \delta \, gatcS \Rightarrow atgc \, \delta \, gatc \ \lambda \\
&= \ atgcdgatc \in L\left(G_p\right)
\end{aligned}
$$

Thus $< atgc, gatc > \ \in \hat{L}\left(G_p\right)$ and $atgc{\cdot}gatc \in \tilde{L}\left(G_p\right)$

S $\Rightarrow$ tS $\Rightarrow$ ttS $\Rightarrow$ ttaS $\Rightarrow$ ttagS $\Rightarrow$ ttagtS $\Rightarrow$ ttagttS $\Rightarrow$ ttagttaS

$\Rightarrow$ ttagtta δ gatcS $\Rightarrow$ ttagtta δ gatccS $\Rightarrow$ ttagtta δ gatccgS

$\Rightarrow$ ttagtta δ gatccgS $\Rightarrow$ ttagtta δ gatccg t S $\Rightarrow$ ttagtta δ gatccgtgS

$\Rightarrow$ ttagtta δ gatccgtg a S $\Rightarrow$ ttagtta δ gatccgtgatS

$\Rightarrow$ ttagtta δ gatccgtgatcS $\Rightarrow$ ttagtta δ gatccgtg a tcgS

$\Rightarrow$ ttagtta δ gatccgtgatcgtS $\Rightarrow$ ttagtta δ gatccgtgatcgttS

$\Rightarrow$ ttagtta δ gatccgtgatcgttaS $\Rightarrow$ ttagtta δ gatccgtgatcgtta δ gatcS

$\Rightarrow$ ttagtta δ gatccgtgatcgtta δ gatctS $\Rightarrow$ ttagtta δ gatccgtgatcgtta δ gatcttS

$\Rightarrow$ ttagtta δ gatccgtgatcgtta δ gatcttgS $\Rightarrow$ ttagtta δ gatccgtgatcgtta δ gatcttgcS

$\Rightarrow$ ttagtta δ gatccgtgatcgtta δ gatcttgcgS $\Rightarrow$ ttagtta δ gatccgtgatcgtta δ gatcttgcg λ

Note that   ttagtta δ gatccgtgatcgtta δ gatcttgcgc $\in L(G_p)$

while   ttagtta, gatccgtgatcgtta, gatcttgcgc $\in \hat{L}(G_p)$

undigested

and   ttagtta· gatccgtgatcgtta· gatcttgcgc $\in \tilde{L}(G_p)$

A new grammar can be devised to obtain complete digests:

$$G_c = \left(\{S\},\{a,c,g,t\},\mathcal{P}_c,S\right) \quad \text{(the "c" subscript refers to a complete digest)}$$

$$\text{where } \mathcal{P}_c = \begin{cases} S & \Rightarrow & gG \mid aS \mid cS \mid tS \mid \delta\,gatcS \mid \lambda \\ G & \Rightarrow & gS \mid aA \mid cS \mid tS \\ A & \Rightarrow & gS \mid aS \mid cS \mid tT \\ T & \Rightarrow & gS \mid aS \mid tS \end{cases}$$

Cut grammars are useful when dealing with context-free and context-sensitive languages and will be discussed in Chapter 5, and Chapter 6.

Assuming that there are no mismatches, abasic pairings, and that only the usual four nucleotide bases A, C, G, and T are involved (in DNA, but this may easily be expanded to RNA by replacing T with U), then we may approach DNA from the viewpoint of a standard FSA, and use a regular Chomsky type 3 grammar as follows.

$$G = \left(\{S\}, \left\{ \begin{pmatrix} A \\ T \end{pmatrix}, \begin{pmatrix} C \\ G \end{pmatrix}, \begin{pmatrix} G \\ C \end{pmatrix}, \begin{pmatrix} T \\ A \end{pmatrix} \right\}, \mathcal{P}, S \right)$$

where: $\mathcal{P} = \left\{ S \Rightarrow \begin{pmatrix} A \\ T \end{pmatrix} S \mid \begin{pmatrix} C \\ G \end{pmatrix} S \mid \begin{pmatrix} G \\ C \end{pmatrix} S \mid \begin{pmatrix} T \\ A \end{pmatrix} S \mid \begin{pmatrix} \lambda \\ \lambda \end{pmatrix} \right\}$

It is understood that $\begin{pmatrix} \lambda \\ \lambda \end{pmatrix}$ does not correspond to any biochemical molecule.

then $L(G) = \left[\begin{pmatrix} A \\ T \end{pmatrix} + \begin{pmatrix} C \\ G \end{pmatrix} + \begin{pmatrix} G \\ C \end{pmatrix} + \begin{pmatrix} T \\ A \end{pmatrix} \right]^*$

It is an easy task to expand the linguistic capability, by adding production rules for non-standard bases, non-standard chelated bases, totally artificial bases, abasic pairs, etc. as discussed in Chapter 3. This can be done for the entire Chomsky hierarchy, as well.

$$S \Rightarrow \begin{pmatrix} isoC \\ isoG \end{pmatrix} S \mid \begin{pmatrix} isoG \\ isoC \end{pmatrix} S \mid \begin{pmatrix} E \\ E \end{pmatrix} S \mid \begin{pmatrix} Dipic \\ Py \end{pmatrix} S \mid \begin{pmatrix} Py \\ Dipic \end{pmatrix} S \mid$$

$$\begin{pmatrix} 5\,MICS \\ A \end{pmatrix} S \mid \begin{pmatrix} A \\ 5\,MICS \end{pmatrix} S \mid \begin{pmatrix} P \\ \varnothing \end{pmatrix} S \mid \begin{pmatrix} \varnothing \\ P \end{pmatrix} S, \text{ etc.}$$

If triple-stranded helices of DNA are to be included, then we would have to consider an expanded grammar once again.

$$G = \left(\{S\}, \left\{ \begin{pmatrix} A \\ T \\ A \end{pmatrix}, \begin{pmatrix} C \\ G \\ C \end{pmatrix}, \begin{pmatrix} G \\ C \\ G \end{pmatrix}, \begin{pmatrix} T \\ A \\ T \end{pmatrix} \right\}, \mathcal{P}, S \right)$$

where: $\mathcal{P} = \left\{ S \Rightarrow \begin{pmatrix} A \\ T \\ A \end{pmatrix} S \mid \begin{pmatrix} C \\ G \\ C \end{pmatrix} S \mid \begin{pmatrix} G \\ C \\ G \end{pmatrix} S \mid \begin{pmatrix} T \\ A \\ T \end{pmatrix} S \mid \begin{pmatrix} \lambda \\ \lambda \\ \lambda \end{pmatrix} \right\}$

However, this would not be very adequate, as triple-stranded DNA often exists in conjunction with double-stranded DNA (see Figure 4.7 as well as Chapter 8, splicing systems). Although a mixed system could be devised, such things as the following (to support mixed helical strings of one, two, or three helices) are not acceptable. Furthermore, there are quadruple-stranded helices of DNA. This will also be discussed in Chapter 5, covering context-free grammars.

RNA Phages, Group I

RNA phages of group I consist of a single-stranded RNA molecule which is 3500 to 4500 nucleotides long, along with a protein coat. At least four proteins are coded on the RNA [24].

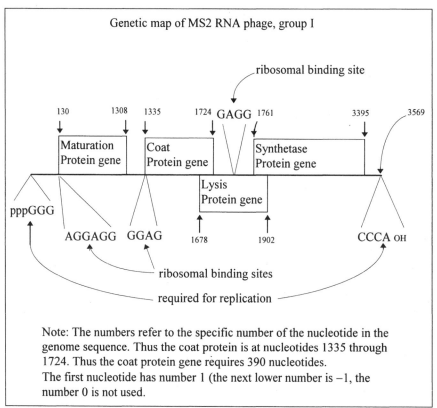

Figure 4.1 Genetic map of MS2 RNA phage, group I: location of bases

The lysis protein gene overlaps the coat protein gene and the synthetase protein gene. A codon frame is every three nucleotides. The lysis protein codon frames are in a different frame than the coat codon frame, by one nucleotide.

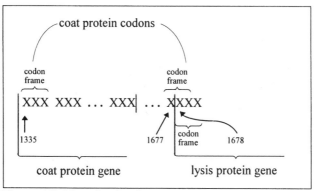

Figure 4.2 Map of codon frames

The maturation protein gene coding region begins with the GUG codon, while the other gene coding regions begin with the AUG codon. The ribosomal binding sites precede the start codons for the maturation, coat, and synthetase genes.

Ribosomal Binding Site (Primary Structure [24, p. 2566])

maturation protein gene	AGGAGG
coat protein gene	GGAG or GGGG
synthetase protein gene	GAGG

Thus a deterministic finite state automaton to accept GGGG or GGAG, or the regular expression GGGG + GGAG = GG (A + G) G follows.

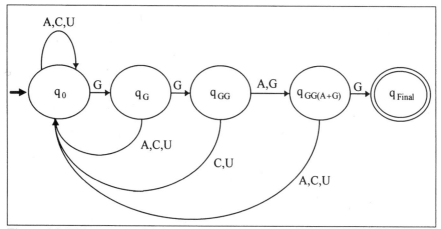

Figure 4.3 FSA to accept GG (A + G) G

A non-deterministic finite state automaton to accept the four ribosomal binding site primary structures follows.

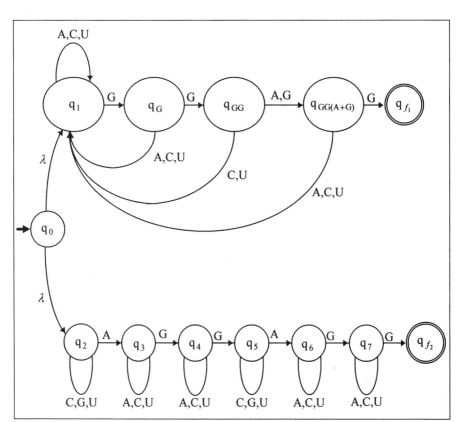

Figure 4.4 Another FSA to accept ribosomal binding site base sequences

It is fairly trivial to convert the above non-deterministic finite state automaton to a deterministic finite state automaton. The deterministic finite state automaton has twenty states, four of which are final (accepting) states.

The Regular (Chomsky Type 3) Language Equivalent to MS2 RNA-Phage, Group I

A series of finite state automata may be constructed and assembled as a non-deterministic FSA, and converted to their corresponding deterministic FSA, which will accept MS2 RNA-Phage, group I. In fact, this deterministic FSA can be converted to its corresponding deterministic finite state transducer, with regular input language, and its corresponding output language of amino acids (translated from the input language codons). A program to do this will follow, in rough outline form.

1. Construct FSA_{A2} which recognizes the ribosomal binding site for the coat gene, with corresponding regular (type 3) language L_{A2}.

2. Construct FSA_{A1} which recognizes the ribosomal binding site for the maturation gene, with corresponding regular (type 3) language L_{A1}.

3. Construct FSA_{A3} which recognizes the ribosomal binding site for the synthetase gene, with corresponding regular (type 3) language L_{A3}.

(Note that FSA complements and complements to their corresponding type 3 regular languages are easily constructed.)

4. Automata for the regions translated by the ribosomes may be constructed, and a partial construction follows [24, p. 2564]; it is transducer $T_{P(0)}$. $T_{P(0)}$ accepts all RNA sequences that:
   a. Have start codon AUG or GUG;
   b. followed by zero or more multiples of three nucleosides (each triple of nucleosides is a codon, and is translated into a corresponding amino acid);
   c. terminated by any stop codon (nucleoside triple that is either UAA, UAG, or UGA).

The corresponding language is $L_1 = L_{A1} \cdot L_{T_{(P0)}} \cdot L_{A2} \cdot L_{T_{(P0)}} \cdot L_{A3} \cdot S^+$, where S^+ is the non-empty regular expression composed of RNA alphabet codons.

$T_{(P0)}$ follows.

T_{P(0)}

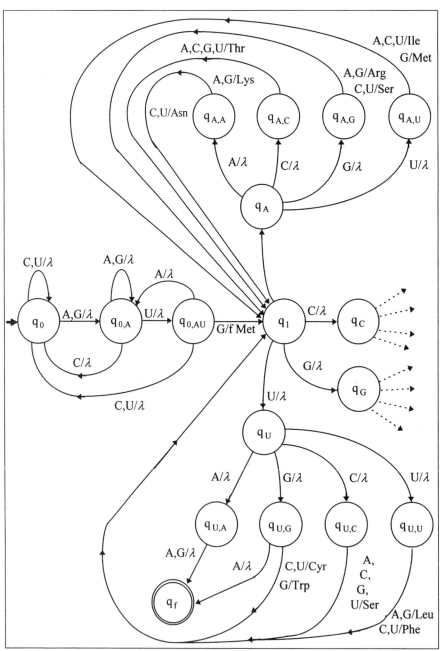

Figure 4.5 FSA for MS2 RNA-Phage, group I

The transducer $T_{P(1)}$ recognizes the overlapping-frame gene for the lysis protein. This transducer looks for the start codon, then proceeds with the frame one letter ahead. Then the transducer seeks a start codon from triplets of the form AUG or GUG. After a triplet start codon is identified, the triples of letters (codons) are transduced into lysis proteins. The overlapping region in the genome is in

$$\left[L_{T_{P(0)}} \cdot L_{A3} \cdot L_{T_{P(0)}} \right] \cap \left[L_{T_{P(1)}} \cdot S^+ \right]$$

Note that finite state automata are a Boolean algebra, thus

$$\left[L_{T_{P(0)}} \cdot L_{A3} \cdot L_{T_{P(0)}} \right] \cap \left[L_{T_{P(1)}} \cdot S^+ \right]$$

is a regular (Chomsky type 3) language also.

$$L_2 = L_{A1} \cdot L_{T_{(P0)}} \cdot L_{A2} \cdot \left(\left[L_{T_{P(0)}} \cdot L_{A3} \cdot L_{T_{P(0)}} \right] \cap \left[L_{T_{P(1)}} \cdot S^+ \right] \right) \cdot S^+ \quad \text{remains a regular}$$

(Chomsky type 3) language. L_2 is a subset of L_1 and includes only those RNA sequences that also code for the lysis gene.

The final consideration is the requirement of phage replication. The replicase function acts as a constraint upon the possible genome terminals. A finite state transducer T_R can be constructed which accepts the end sequence ACCC seeking the starting sequence GGG. In this case, the Chomsky type 3 (regular) language $L_{T_R}^r$ will correspond to finite state transducer T_R, but in reverse order. We thus obtain $L_3 = L_2 \cap L_{T_R}^r$. Note that regular language L_3 models the genome for MS2 RNA phage, group I.

The transducer $T_{P(1)}$ follows.

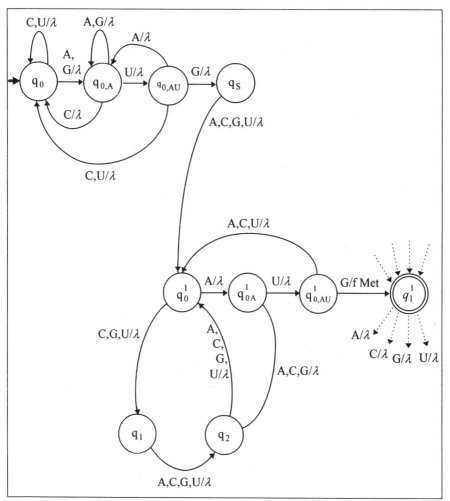

Figure 4.6 Transducer $T_{P(1)}$ for overlapping-frame gene for the lysis protein

$$V_T = \left\{ \begin{matrix} \begin{pmatrix} A \\ T \\ A \end{pmatrix}, \begin{pmatrix} C \\ G \\ C \end{pmatrix}, \begin{pmatrix} G \\ C \\ G \end{pmatrix}, \begin{pmatrix} T \\ A \\ T \end{pmatrix}, \begin{pmatrix} A \\ T \\ \lambda \end{pmatrix}, \begin{pmatrix} A \\ \lambda \\ T \end{pmatrix}, \begin{pmatrix} \lambda \\ A \\ T \end{pmatrix}, \begin{pmatrix} C \\ G \\ \lambda \end{pmatrix}, \begin{pmatrix} C \\ \lambda \\ G \end{pmatrix}, \begin{pmatrix} \lambda \\ C \\ G \end{pmatrix}, \begin{pmatrix} G \\ C \\ \lambda \end{pmatrix}, \begin{pmatrix} G \\ \lambda \\ C \end{pmatrix}, \begin{pmatrix} \lambda \\ G \\ C \end{pmatrix} \\ \begin{pmatrix} T \\ A \\ \lambda \end{pmatrix}, \begin{pmatrix} T \\ \lambda \\ A \end{pmatrix}, \begin{pmatrix} \lambda \\ T \\ A \end{pmatrix}, \begin{pmatrix} A \\ \lambda \\ \lambda \end{pmatrix}, \begin{pmatrix} \lambda \\ A \\ \lambda \end{pmatrix}, \begin{pmatrix} \lambda \\ \lambda \\ A \end{pmatrix}, \begin{pmatrix} C \\ \lambda \\ \lambda \end{pmatrix}, \begin{pmatrix} \lambda \\ C \\ \lambda \end{pmatrix}, \begin{pmatrix} \lambda \\ \lambda \\ C \end{pmatrix}, \begin{pmatrix} G \\ \lambda \\ \lambda \end{pmatrix}, \begin{pmatrix} \lambda \\ G \\ \lambda \end{pmatrix}, \begin{pmatrix} \lambda \\ \lambda \\ G \end{pmatrix}, \begin{pmatrix} T \\ \lambda \\ \lambda \end{pmatrix}, \begin{pmatrix} \lambda \\ T \\ \lambda \end{pmatrix}, \begin{pmatrix} \lambda \\ \lambda \\ T \end{pmatrix}, \begin{pmatrix} \lambda \\ \lambda \\ \lambda \end{pmatrix} \end{matrix} \right\}$$

Some of the triple helix possibilities are discussed in [30, p. 155], with mismatches discussed in [30, p. 138].

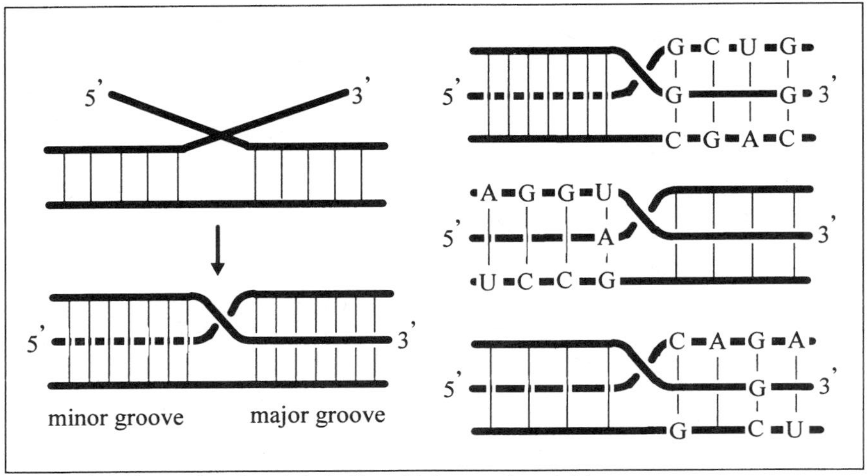

Figure 4.7 Example of a triple helix with base mismatching

In Vitro Molecular Computation

The purpose of this book is to examine how aspects of automata theory may be used to model various phenomena in a number of areas outside computer science, electrical engineering, or mathematics. It is well known that there is a one-to-one functional relationship or parallel between computational machinery and the linguistic hierarchy of Chomsky (see the diagrammatic representation of this parallel, below). However, another effort is underway, namely, the actual and practical construction of computational machinery based upon new principles of design. Specifically, the construction of computational machinery not using transistors, direct-current power supplies, digital circuits, arithmetic/logic units, memory modules, etc., but rather, constructed from biologically derived components, such as DNA macro-molecules, ATP, or artificial molecules to provide power based upon photosynthesis, etc. or hybrids of traditional components with biologically derived components, even living components.[b] This program is often referred to as "molecular computation." Such a program seems far off in the future and lies mostly outside the purpose of this book. However, restricting ourselves to the purpose of this book, namely, how aspects of automata theory may be used to model various phenomena, a short discussion of aspects of automata theory in relation to molecular computation is relevant to our objectives. We have in fact already encountered in vitro and in vivo applications in Chapter 3, and we will see another reference in chapter 7 (a universal Turing machine).

Primitive Components in Molecular Computation

As most of the devices in this area lie outside the area of automata theory that is described in this book, little attention will be given to this area, other than mentioning some recent achievements.

A DNA-based computational device has been developed to accomplish addition. This is discussed in [55].

A DNA-based computational device has been developed to accomplish multiplication. This is discussed in [116].

A method of solving Hamiltonian graph problems has been discussed in [2].

---

[b] It is not the purpose of this book to examine moral questions that might arise in the context of mixing or using biologically derived or living components in the devices used or to be used in molecular computation. However, the question as to the morality of this program ought not to be ignored or considered insignificant.

Finite State Automata

Two methods are proposed [57]; a method based upon ligation and a method without ligation.

Sample FSA:

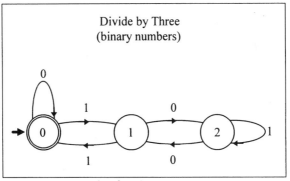

Figure 4.8 Example of a FSA

Samples:

$$\frac{011}{011} = 1 + \frac{0}{011}$$

$$\frac{100}{011} = 1 + \frac{1}{011}$$

$$\frac{101}{011} = 1 + \frac{10}{011}$$

$$\frac{101110}{011} = 1111 + \frac{1}{011}$$

Ligation Method

The approach taken is as follows.

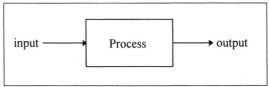

Figure 4.9 How the experiment is carried out

where:

input: problem is encoded in oligonucleotides,[c] encoding alphabet is A, C, G, T
processing: hybridization of DNA molecules and ligation of DNA
output: extraction of processed DNA.

Note: Hybridization is the joining of two complementary DNA strands into a double helix.

Ligation is the joining of two DNA double-helix molecules into one DNA double helix.

Recall that restriction enzymes are proteins that cut the DNA helix at specific sites. Two such restriction enzyme sites are as follows.

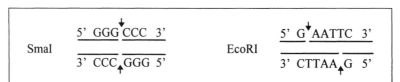

Figure 4.10 Examples of restriction enzymes that may be used in this experiment

Input strings are called adapters, that are a double-stranded DNA helix, usually with a single-stranded overhang (a sticky end). The double-stranded portion encodes an input (a 0 or 1 in our example), while the overhang encodes the present or current state. Thus the adapter is a physical realization of the mapping $\delta(\text{state}, \text{input})$.

---

[c] Oligonucleotides are written in the 3' to 5' direction [57].

Example:

State	nucleotide encoding
0	ttat
1	gctg
2	ctca

In this specific case, input "0" corresponds to GAG, and input "1" corresponds to GCA.

Thus an adapter is a molecule that encodes an input, for a specific state:

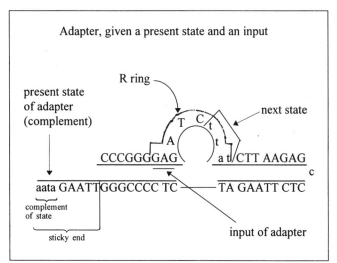

Figure 4.11 Representation of an adapter, with states and R ring

Note: R rings or R loops are also discussed in the next chapter.

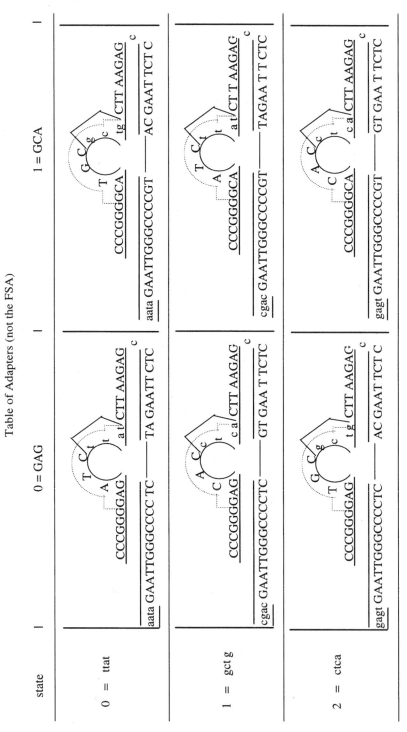

Table of Adapters (not the FSA)

Figure 4.12

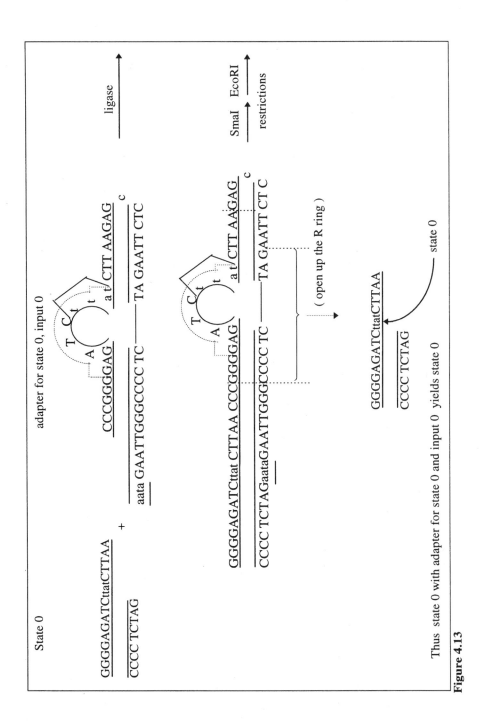

Figure 4.13

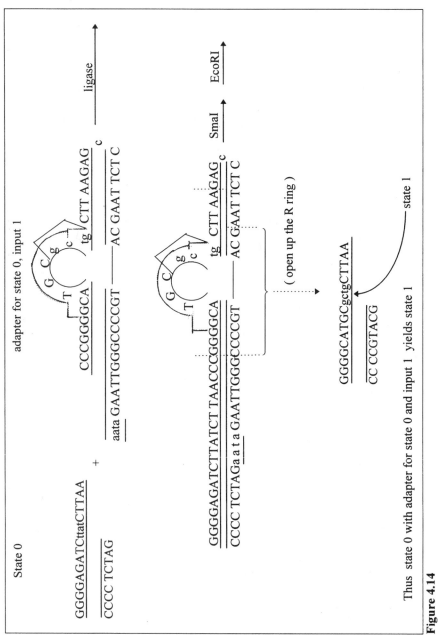

Figure 4.14

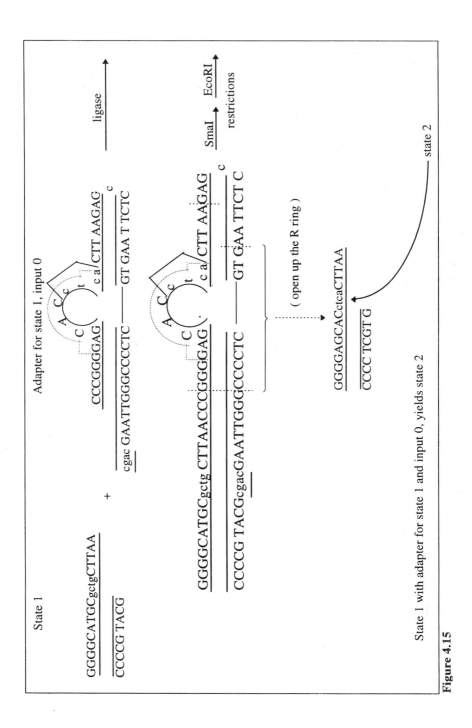

Figure 4.15

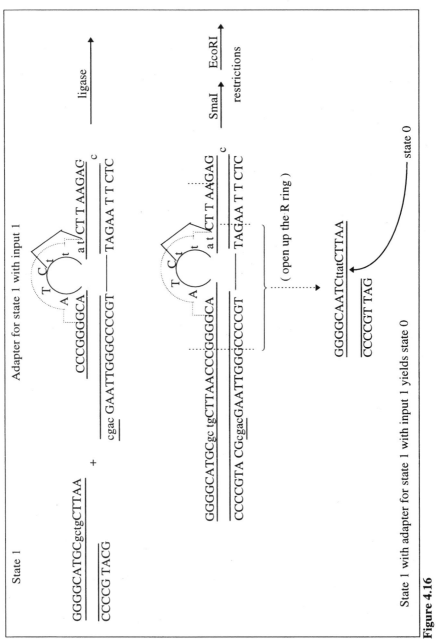

Figure 4.16

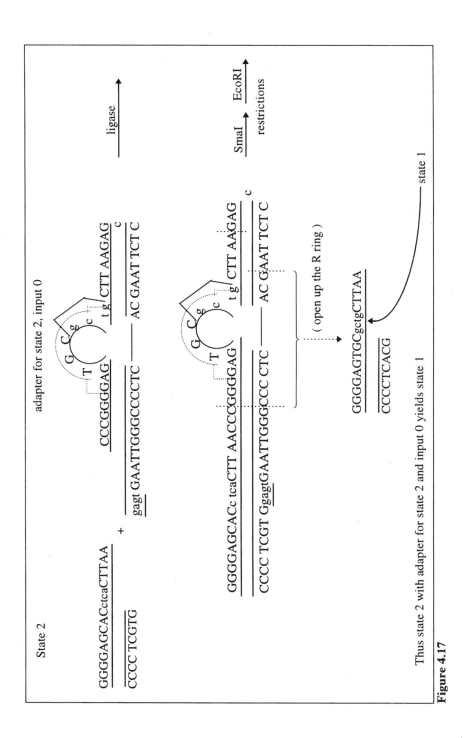

Figure 4.17

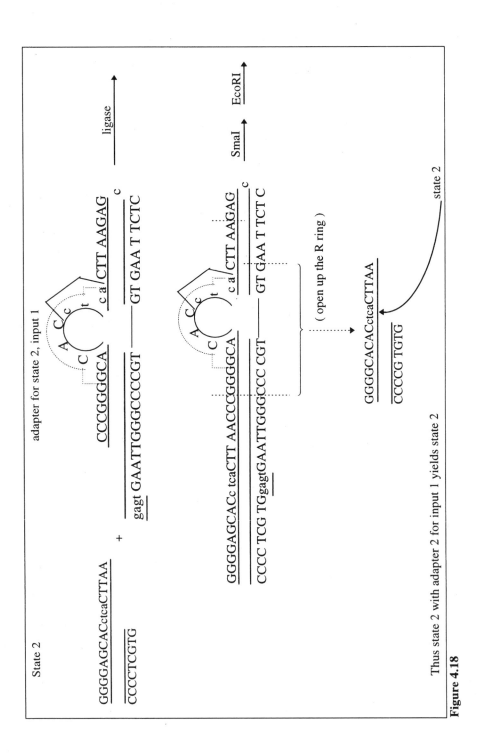

Figure 4.18

To nicely display the FSA, we note the following.

$\delta(\text{state, input}) = \delta(\text{state, adapter[state, input]})$, and we obtain the desired FSA:

state	0	1
0	0	1
1	2	0
2	1	2

Showing the adapters, then we obtain:

state	0	1
0	$\delta(0, \text{adapter}[0,0])$	$\delta(0, \text{adapter}[0,1])$
1	$\delta(1, \text{adapter}[1,0])$	$\delta(1, \text{adapter}[1,1])$
2	$\delta(2, \text{adapter}[2,0])$	$\delta(2, \text{adapter}[2,1])$

Figure 4.19 FSA (using the adapters)

Alternatively,

$$\delta(0, \text{adapter}[0,0]) = 0, \qquad \delta(0, \text{adapter}[0,1]) = 1$$
$$\delta(1, \text{adapter}[1,0]) = 2, \qquad \delta(1, \text{adapter}[1,1]) = 0$$
$$\delta(2, \text{adapter}[2,0]) = 1, \qquad \delta(2, \text{adapter}[2,1]) = 2$$

Non-Deterministic Finite State Automata

Before we proceed to the method of implementation that does not utilize ligation, the nondeterministic finite state automaton will be considered.

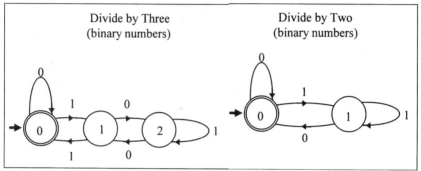

Figure 4.20 Examples of deterministic FSA

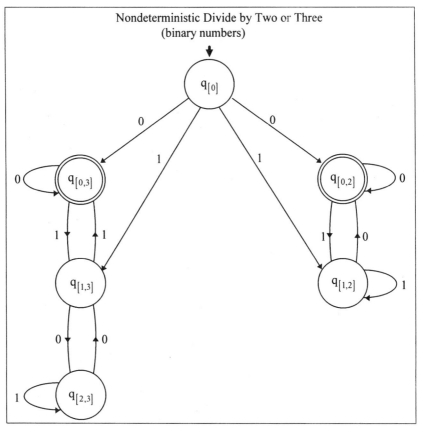

Figure 4.21 Example of a NDFSA, $|\delta(q_{(0)}, 0)| > 1$, and $|\delta(q_{(0)}, 1)| > 1$

state	nucleotide sequence
$q_{[0]}$	atat
$q_{\lfloor 0,3 \rfloor}$	ttat
$q_{\lfloor 1,3 \rfloor}$	gctg
$q_{\lfloor 2,3 \rfloor}$	ctca
$q_{\lfloor 0,2 \rfloor}$	tatt
$q_{\lfloor 1,2 \rfloor}$	gtcg

Figure 4.22 Representation of states

From this point forward, the method is the same as in the previous deterministic case [58, pp. 64-70].

Implementation Without Ligation

There are distinct advantages to the non-ligation method: the molecules are reusable [57, p. 482], and the process can run in uninterrupted fashion. A library of transition molecules is created. These molecules encode the input set and state representations [57, p. 482]. Transitions take place by Watson-Crick hybridization [57, p. 482] where hybridization replaces ligation [57, p. 483]. The input molecule is a concatenation of codeword submolecules in the 5´ to 3´ direction. There is a specific starter code word at the 5´ end, and a terminating codeword at the 3´ end [57, p. 483]. Molecules representing the terminal state that has been attained may be detected optically (these molecules may be made to have a fluorescence property), thus easing their detection [57, p. 483].

The molecules used to implement this method will now be described [57, pp. 482-484, 486, 487]; see Figures 4.23 through 4.25.

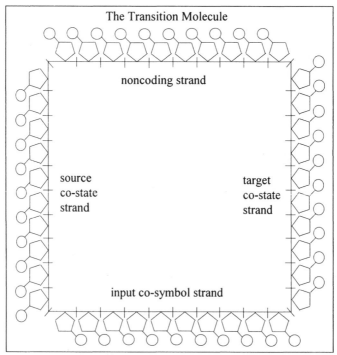

The Transition Molecule

noncoding strand

source
co-state
strand

target
co-state
strand

input co-symbol strand

Figure 4.23 Ring of bases (without ligation)

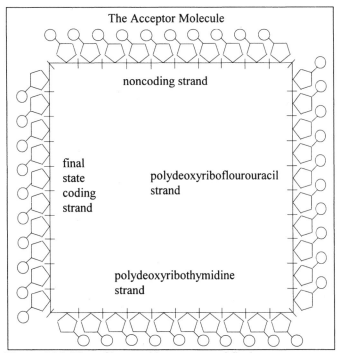

Figure 4.24 Ring of bases with acceptor and final state

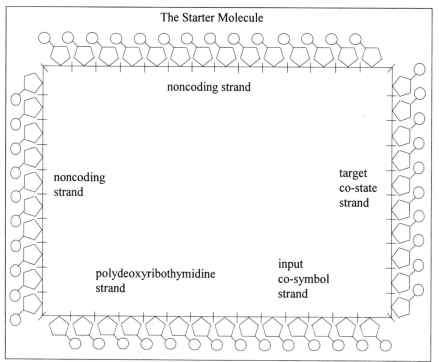

Figure 4.25 Ring of bases with starting input

Chapter 5: Context-Free Languages: DNA and RNA

Complexion

"The term 'complexio' [231, pp. 101, 102] was, from the twelfth century, the Latin commonly used for the Greek *crasis*, or temperament, that is to say, the balance of the qualities of hot, wet, cold, and dry resulting from the mixture of the elements in the human body. Since it served as a fundamental concept, not only in physiology but also in pathology and therapy, complexion theory provided important support for the idea that medicine constituted a unified and rational body of knowledge. The general theory, already quite fully developed in the works of Galen, underwent considerable further elaboration during the Middle Ages.

"Complexion also varied among different peoples or geographic regions; Sythians, who lived in a cold climate, were supposed to be colder and moister in complexion than Ethiopians, who lived under the hot sun."

The Humors

"The concept of humors [231, pp. 104-106]–that is, specific bodily fluids essential to the physiological functioning of the organism–originated at a very early stage of Greek medicine. Various Hippocratic treatises mention one or more of the fluids; *On the Nature of Man* presents what was to become the standard set of four: blood, phlegm, bile (also termed choler, or red or yellow bile), and black bile (or melancholy). The theory of the humors probably developed because bodily fluids of all kinds played a large part in ancient, and subsequently medieval and Renaissance, physiology, diagnosis, and therapy.

"Like all body parts, they (the humors) were themselves complexionate.

"Hence, the balance of humors was held to be responsible for physiological as well as physical disposition, a belief enshrined in the survival of the English adjectives sanguine, phlegmatic, choleric, and melancholy to describe traits of character.

"Medical theory asserted that the human body exists in either health, sickness, or a neutral state between the two. Deviations from health were classified into congenital malformations (in medieval Latin, *mala compositio* of the body), complexional imbalance (*mala complexio*), and trauma (*solutio continuitatis*, or break in the body's continuity). This classification placed almost all internal illness in the domain of complexional imbalance. Relatively little attention was paid to the first of these three categories, and when surgery emerged from medicine as a separate occupation and discipline in the West during the twelfth and thirteenth centuries, the management of trauma became the characteristic task of the surgeon."

Palindromes

A special example of context-free languages are the languages composed of palindromes. Palindromes are of special importance in their relationship to DNA and RNA. Before going any further, a good idea of what a palindrome is can be obtained through examining examples of palindromes.

Definition: A palindrome over V_T is words or strings composed of terminal symbols in V_T but where the strings are spelled the same way in either direction.

Some examples follow.

Example 1:

$$\text{desserts stressed}$$

$$\xrightarrow{\hspace{4cm}} \Big\} \quad \text{same spelling, either direction}$$

Note that in example 1, $V_T = \{d, e, s, r, t, \Delta\}$ where Δ signifies a "blank" or "space".

Example 2:

$$\text{able was I ere I saw elba}$$

$$\xrightarrow{\hspace{4cm}} \Big\} \quad \text{same spelling, either direction}$$

Example 3:

$$N\,I\,Y\,O\,N\,A\,N\,O\,M\,H\,M\,A\,T\,A\,M\,H\,M\,O\,N\,A\,N\,O\,Y\,I\,N$$

same spelling, either direction

(Greek inscription on the sacred font in the courtyard of the Hagia Sophia in Constantinople. The inscription means "wash your sins not only your face")

Example 4:

$$\text{abbcccdcccbba}$$

$$\xrightarrow{\hspace{2cm}} \Big\} \quad \text{same spelling, either direction}$$

Example 5:

$$\left\{ w \cdot w^R \mid w \in V_T^* \right\}$$

$$\rightleftarrows \Big\} \quad \text{same spelling, either direction}$$

Example 6:

each of the three words: racecar, kayak, level

Context-Free Grammars: DNA and RNA

Constructing context-free grammars that can generate palindromes is an easy task. To obtain the palindrome "desserts stressed", we see that $V_T = \{$ d, e, r, s, t, Δ $\}$ (where Δ will signify a space character). Note that the palindrome has an odd number of characters. The grammar follows.

$$G^{pal} = \left(\{S\}, \{d, e, r, s, t, \Delta\}, \mathcal{P}^{pal}, S \right) \text{ where:}$$

$$\mathcal{P}^{pal} = \left\{ S \Rightarrow dSd \mid eSe \mid rSr \mid sSs \mid tSt \mid \Delta \right\}$$

A sample derivation:

$$S \Rightarrow dSd \Rightarrow deSed \Rightarrow desSsed \Rightarrow dessSssed \Rightarrow desseSessed$$
$$\Rightarrow desserSressed \Rightarrow dessertStressed \Rightarrow dessertsSstressed$$
$$\Rightarrow desserts \Delta stressed$$

It is an easy task to construct context-free grammars that can generate palindrome-like languages for DNA and RNA.

$$G^{DNA_1} = \left(\{S\}, \{a, c, g, t\} \cup \{\bar{a}, \bar{c}, \bar{g}, \bar{t}\}, \mathcal{P}^{DNA_1}, S \right)$$
$$\text{where: } \mathcal{P}^{DNA_1} = \left\{ S \Rightarrow aS\bar{a} \mid cS\bar{c} \mid gS\bar{g} \mid tS\bar{t} \mid \lambda \right\}$$

However, using the Watson-Crick complementary relationship, we know that the following is true.

$$\bar{a} = t \qquad \bar{c} = g$$
$$\bar{t} = a \qquad \bar{g} = c$$

Thus a better grammar for DNA is the following.

$$G^{DNA} = \left(\{S\}, \{a, c, g, t\}, \mathcal{P}^{DNA}, S \right)$$
$$\text{where: } \mathcal{P}^{DNA} = \left\{ S \Rightarrow aSt \mid cSg \mid gSc \mid tSa \mid \lambda \right\}$$

The question could be asked, "How is the language $L_{DNA} = L\left(G^{DNA}\right)$ composed of palindromes?"

Let us examine a derivation in $L_{DNA_1} = L\!\left(G^{DNA_1}\right)$; then we will examine the corresponding string in $L_{DNA} = L\!\left(G^{DNA}\right)$.

In $L_{DNA_1} = L\!\left(G^{DNA_1}\right)$,

$$
\begin{aligned}
S &\Rightarrow t\,S\,\bar{t} \Rightarrow t\,a\,S\,\bar{a}\,\bar{t} \Rightarrow t\,a\,t\,S\,\bar{t}\,\bar{a}\,\bar{t} \Rightarrow t\,a\,t\,a\,S\,\bar{a}\,\bar{t}\,\bar{a}\,\bar{t} \\
&\Rightarrow t\,a\,t\,a\,c\,S\,\bar{c}\,\bar{a}\,\bar{t}\,\bar{a}\,\bar{t} \Rightarrow t\,a\,t\,a\,c\,g\,S\,\bar{g}\,\bar{c}\,\bar{a}\,\bar{t}\,\bar{a}\,\bar{t} \\
&\Rightarrow t\,a\,t\,a\,c\,g\,t\,S\,\bar{t}\,\bar{g}\,\bar{c}\,\bar{a}\,\bar{t}\,\bar{a}\,\bar{t} \\
&\Rightarrow t\,a\,t\,a\,c\,g\,t\,a\,S\,\bar{a}\,\bar{t}\,\bar{g}\,\bar{c}\,\bar{a}\,\bar{t}\,\bar{a}\,\bar{t} \\
&\Rightarrow t\,a\,t\,a\,c\,g\,t\,a\,\lambda\,\bar{a}\,\bar{t}\,\bar{g}\,\bar{c}\,\bar{a}\,\bar{t}\,\bar{a}\,\bar{t} \\
&= t\,a\,t\,a\,c\,g\,t\,a\,\cdot\,\bar{a}\,\bar{t}\,\bar{g}\,\bar{c}\,\bar{a}\,\bar{t}\,\bar{a}\,\bar{t}
\end{aligned}
$$

While the derived string is not a palindrome, it is palindromic in an obvious sense.

Let us examine the string that would have been derived in $L_{DNA} = L\!\left(G^{DNA}\right)$:

$$t\,a\,t\,a\,c\,g\,t\,a\,\cdot\,t\,a\,c\,g\,t\,a\,t\,a$$

but if we view this as a double-stranded DNA helix:

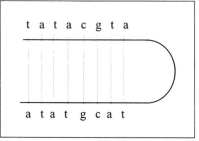

Figure 5.1 RNA folded into a doble-helix like DNA

RNA is typically single-stranded, but the same energy-stabilizing forces that act on a double-stranded DNA helix cause RNA to curl itself into a double helix, thus RNA is also palindromic, like DNA. The big difference is that in RNA, "t" is replaced with "u". Review the discussion in Chapter 1.

$$
G^{RNA_1} = \left(\{S\}\,,\{a,\ c,\ g,\ u\} \cup \{\bar{a},\ \bar{c},\ \bar{g},\ \bar{u}\},\ \mathcal{P}^{RNA_1},\ S\right)
$$
$$
\text{where:}\quad \mathcal{P}^{RNA_1} = \left\{S \Rightarrow aS\bar{a}\ \mid\ cS\bar{c}\ \mid\ gS\bar{g}\ \mid\ uS\bar{u}\ \mid\ \lambda\right\}
$$

Similarly, using the Watson-Crick complementary relationship (but for RNA, now), we know that the following is true (sometimes g and u are paired).

$$\overline{a} = u \qquad \overline{c} = g$$
$$\overline{u} = a \qquad \overline{g} = c$$

Thus a better grammar for RNA is the following [160, pp. 586-588].

$$G^{RNA} = \left(\{S\}, \{a, c, g, u\}, \mathcal{P}^{RNA}, S\right)$$
$$\text{where: } \mathcal{P}^{RNA} = \left\{S \Rightarrow aSu \mid cSg \mid gSc \mid uSa \mid \lambda\right\}$$

Both DNA and RNA may have recursive instances of palindromic structures. However, a grammar supporting recursive or repeated palindromic structures is still context-free. Examples of recursive or repeated palindromes:

1. repeated palindromes: desserts stressed · eve
2. recursive palindromes: desserts eve stressed

DNA and RNA grammars that support repeated and recursive palindromic structures [160, p. 589]:

$$G^{DNA_r} = \left(\{S\}, \{a, c, g, t\}, \mathcal{P}^{DNA_r}, S\right)$$
$$\text{where: } \mathcal{P}^{DNA_r} = \left\{S \Rightarrow SS \mid aSt \mid cSg \mid gSc \mid tSa \mid \lambda\right\}$$

$$G^{RNA_r} = \left(\{S\}, \{a, c, g, u\}, \mathcal{P}^{RNA_r}, S\right)$$
$$\text{where: } \mathcal{P}^{RNA_r} = \left\{S \Rightarrow SS \mid aSu \mid cSg \mid gSc \mid uSa \mid \lambda\right\}$$

More complicated structures, such as supporting copies of substrings, would require a grammar more complex than context-free, such as context-sensitive. This will be discussed at length in the next chapter.

Dyck Languages and the Chomsky-Schützenberger Theorem

If the grammars for DNA and RNA are viewed palindromically, an interesting fact emerges.

$$G^{DNA_2} = \left(\{S\}, \ \{a, \ c, \ g, \ t\} \ \cup \ \{\bar{a}, \ \bar{c}, \ \bar{g}, \ \bar{t}\}, \ \mathcal{P}^{DNA_2}, \ S \right)$$

$$= \left(\{S\}, \ \{a, \ c, \ g, \ t\}, \ \mathcal{P}^{DNA_2}, \ S \right)$$

$$\text{where:} \quad \mathcal{P}^{DNA_2} = \left\{ S \ \Rightarrow \ SS \mid aS\bar{a} \mid cS\bar{c} \mid gS\bar{g} \mid tS\bar{t} \mid \lambda \right\}$$

and

$$G^{RNA_2} = \left(\{S\}, \ \{a, \ c, \ g, \ u\} \ \cup \ \{\bar{a}, \ \bar{c}, \ \bar{g}, \ \bar{u}\}, \ \mathcal{P}^{RNA_2}, \ S \right)$$

$$\text{where:} \quad \mathcal{P}^{RNA_2} = \left\{ S \ \Rightarrow \ SS \mid aS\bar{a} \mid cS\bar{c} \mid gS\bar{g} \mid uS\bar{u} \mid \lambda \right\}$$

The interesting fact is that both these grammars are effectively Dyck grammars!

The Chomsky-Schützenberger Theorem states:

$L = h[D \cap R]$, where L is an arbitrary context-free language, h is a homomorphism, D is a Dyck language, and R is a regular (type 3) language.

Thus arbitrary context-free languages to describe DNA and RNA, may be used in the Chomsky-Schützenberger Theorem if G^{DNA_2} or G^{RNA_2} are used as the Dyck languages [158, pp. 104, 105].

Thus note the comment: "The language of genes really represents the intersection of separate languages for transcription, processing, translation and even for the encoded protein sequence, which must specify the folded structure of the protein with its rich array of nested and crossing dependencies" [160, p. 591].

Splicing System-Like View

A small diversion (promised in the Chapter 4, when discussing type 3 grammars and FSA) would be à propos. We have just studied how DNA and RNA might be viewed as context-free grammars. However, it is very interesting that the context-free grammars to describe DNA and RNA may also be viewed as type 3 grammars, if the terminal sets are more complex. Thus a review of the discussion under regular grammars is timely.

Assuming that there are no mismatches, abasic pairings, and that only the usual four nucleotide bases A, C, G, and T are involved (in DNA, but this may easily be expanded to RNA by replacing T with U), then we may approach DNA from the viewpoint of a standard FSA, and use a regular type 3 grammar as follows.

$$G = \left(\{S\}, \left\{ \binom{A}{T}, \binom{C}{G}, \binom{G}{C}, \binom{T}{A} \right\}, \mathcal{P}, S \right)$$

$$\text{where: } \mathcal{P} = \left\{ S \Rightarrow \binom{A}{T} S \mid \binom{C}{G} S \mid \binom{G}{C} S \mid \binom{T}{A} S \mid \binom{\lambda}{\lambda} \right\}$$

It is understood that $\binom{\lambda}{\lambda}$ does not correspond to any biochemical molecule.

$$\text{Then } L(G) = \left[\binom{A}{T} + \binom{C}{G} + \binom{G}{C} + \binom{T}{A} \right]^{*}$$

Thus our previous generation of a DNA double-helix can be accomplished as follows.

$$S \Rightarrow \binom{T}{A} S \Rightarrow \binom{T}{A}\binom{A}{T} S \Rightarrow \binom{T}{A}\binom{A}{T}\binom{T}{A} S \Rightarrow \binom{T}{A}\binom{A}{T}\binom{T}{A}\binom{A}{T} S$$

$$\Rightarrow \binom{T}{A}\binom{A}{T}\binom{T}{A}\binom{A}{T}\binom{C}{G} S \Rightarrow \binom{T}{A}\binom{A}{T}\binom{T}{A}\binom{A}{T}\binom{C}{G}\binom{G}{C} S$$

$$\Rightarrow \binom{T}{A}\binom{A}{T}\binom{T}{A}\binom{A}{T}\binom{C}{G}\binom{G}{C}\binom{T}{A} S \Rightarrow \binom{T}{A}\binom{A}{T}\binom{T}{A}\binom{A}{T}\binom{C}{G}\binom{G}{C}\binom{T}{A}\binom{A}{T} S$$

$$\Rightarrow \binom{T}{A}\binom{A}{T}\binom{T}{A}\binom{A}{T}\binom{C}{G}\binom{G}{C}\binom{T}{A}\binom{A}{T}\binom{\lambda}{\lambda} = \binom{T}{A}\binom{A}{T}\binom{T}{A}\binom{A}{T}\binom{C}{G}\binom{G}{C}\binom{T}{A}\binom{A}{T}$$

This is a well-known occurrence in the theory of computation. It has been noted in a paper by Siromoney, that the language $L = \{ a^n b^n c^n \mid n \geq 0 \}$ usually viewed as context-sensitive, may be viewed as context-free (if a matrix grammar is used). Changing the theoretical basis (even only V_T) supports powerful changes in the theoretical viewpoint.

The above ideas can be extended to include ideas discussed under splicing systems, and domino systems in Chapter 8, comprehending restriction endonucleases. It is impossible to examine all the possible extensions, but to get an idea of some of the possibilities, let us consider the following example.

Let us recall two common restriction endonucleases, BclI and MboI.

Figure 5.2 Restriction endonuclease cleavage

Let us construct the context-free grammar G^{MB} (once again, we assume no mismatches, triple helices, abasic pairs, etc.).

$$G^{MB} = \left(\{S, \ B, \ M\}, \ V_T, \ \mathcal{P}^{MB}, \ S \right)$$

where:

$$V_T = \left\{ \begin{pmatrix} T \\ ACTAG \end{pmatrix}, \begin{pmatrix} GATCA \\ T \end{pmatrix}, \begin{pmatrix} GAG \\ CTC \end{pmatrix}, \begin{pmatrix} CGG \\ GCC \end{pmatrix}, \begin{pmatrix} A \\ T \end{pmatrix}, \begin{pmatrix} C \\ G \end{pmatrix}, \begin{pmatrix} G \\ C \end{pmatrix}, \begin{pmatrix} T \\ A \end{pmatrix}, \begin{pmatrix} \lambda \\ \lambda \end{pmatrix} \right\}$$

and the production rules $\mathcal{P}^{MB}$ are

$$S \ \Rightarrow \ S\begin{pmatrix} T \\ ACTAG \end{pmatrix} B \ | \ S\begin{pmatrix} GAG \\ CTC \end{pmatrix} M \ | \ \begin{pmatrix} A \\ T \end{pmatrix} S \ | \ \begin{pmatrix} C \\ G \end{pmatrix} S \ | \ \begin{pmatrix} G \\ C \end{pmatrix} S \ | \ \begin{pmatrix} T \\ A \end{pmatrix} S \ | \ \begin{pmatrix} l \\ l \end{pmatrix}$$

$$B \ \Rightarrow \ \begin{pmatrix} GATCA \\ T \end{pmatrix} S$$

$$M \ \Rightarrow \ \begin{pmatrix} CGG \\ GCC \end{pmatrix} S$$

Thus a sample derivation:

$$S \Rightarrow S\binom{T}{ACTAG}B \Rightarrow S\binom{T}{ACTAG}\binom{GATCA}{T}S \Rightarrow S\binom{T}{ACTAG}\binom{GATCA}{T}\binom{G}{C}S$$

$$\Rightarrow S\binom{T}{ACTAG}\binom{GATCA}{T}\binom{G}{C}\binom{T}{A}S \Rightarrow S\binom{T}{ACTAG}\binom{GATCA}{T}\binom{G}{C}\binom{T}{A}\binom{\lambda}{\lambda}$$

$$\Rightarrow \binom{C}{G}S\binom{T}{ACTAG}\binom{GATCA}{T}\binom{G}{C}\binom{T}{A}\binom{\lambda}{\lambda}$$

$$\Rightarrow \binom{C}{G}\binom{GAG}{CTC}M\binom{T}{ACTAG}\binom{GATCA}{T}\binom{G}{C}\binom{T}{A}\binom{\lambda}{\lambda}$$

$$\Rightarrow \binom{C}{G}\binom{GAG}{CTC}\binom{CGG}{GCC}S\binom{T}{ACTAG}\binom{GATCA}{T}\binom{G}{C}\binom{T}{A}\binom{\lambda}{\lambda}$$

$$\Rightarrow \binom{C}{G}\binom{GAG}{CTC}\binom{CGG}{GCC}\binom{A}{T}S\binom{T}{ACTAG}\binom{GATCA}{T}\binom{G}{C}\binom{T}{A}\binom{\lambda}{\lambda}$$

$$\Rightarrow \binom{C}{G}\binom{GAG}{CTC}\binom{CGG}{GCC}\binom{A}{T}\binom{T}{A}S\binom{T}{ACTAG}\binom{GATCA}{T}\binom{G}{C}\binom{T}{A}\binom{\lambda}{\lambda}$$

$$\Rightarrow \binom{C}{G}\binom{GAG}{CTC}\binom{CGG}{GCC}\binom{A}{T}\binom{T}{A}\binom{\lambda}{\lambda}\binom{T}{ACTAG}\binom{GATCA}{T}\binom{G}{C}\binom{T}{A}\binom{\lambda}{\lambda}$$

$$= \binom{CGAGCGGATTGATCAGT}{GCTCGCCTAACTAGTCA}$$

Similar to the discussion in the previous chapter, we could modify this language by supporting non-standard bases, non-standard chelated bases, totally artificial bases, abasic pairs, etc., as discussed in Chapter 3. We would have to add the following to V_T,

$$\binom{\kappa}{\pi}, \binom{\pi}{\kappa}, \binom{H}{H}, \binom{Dipic}{Py}, \binom{Py}{Dipic}, \binom{MICS}{T}, \binom{T}{MICS}, \binom{P}{\varnothing}, \binom{\varnothing}{P}, \text{etc.}$$

and in addition, we would have to add the following to the production rules:

$$S \Rightarrow \binom{\kappa}{\pi}S \mid \binom{\pi}{\kappa}S \mid \binom{H}{H}S \mid \binom{Dipic}{Py}S \mid \binom{Py}{Dipic}S \mid$$

$$\binom{MICS}{T}S \mid \binom{T}{MICS}S \mid \binom{P}{\varnothing}S \mid \binom{\varnothing}{P}S, \text{etc.}$$

Our little diversion is over; we end our discussion of context-free grammars that deal with methods similar to splicing systems (splicing systems will be discussed at greater length in Chapter 8).

Cut Grammars

Recall that cut grammars [164] were discussed in a previous chapter limited to Chomsky type 3 regular grammars, and finite state automata (see Chapter 4). At that time, discussion of cut grammars relating to context-free grammars was postponed. We shall now resume the discussion of context-free cut grammars, and further study takes place in the next chapter.

As a review:

Definition of a cut grammar:

$$G = \left(V_N,\ V_T,\ \mathcal{P},\ S\right)$$ where the production rules $\mathcal{P}$ are defined as follows:

$$\left(V_N \cup V_T\right)^* V_N \left(V_N \cup V_T\right)^* \times \left(V_N \cup V_T \cup \{\delta\}\right)^* \text{ and } \delta \notin V_N \cup V_T$$

Note: as $\delta \notin \left(V_N \cup V_T\right)^* V_N \left(V_N \cup V_T\right)^* \times \left(V_N \cup V_T \cup \{\delta\}\right)^*$, then δ does not appear in any context (left-hand side of any production rule)

Definition of a cut language:

$$L(G) = \left\{ \omega \mid \omega \in \left[V_T \cup \{\delta\}\right]^* \& \left(S \overset{*}{\Rightarrow} \omega\right) \right\}$$

Definition: Given that ω is a string in a cut language and $\omega = \omega_1 \delta \omega_2 \delta \ \ldots \ \omega_n \delta$ and $\omega_i \in V_T^*$

then

1. $\hat{\omega} = \left\langle \omega_1, \omega_2, \ldots, \omega_n \right\rangle$ an ordered tuple of strings with all instaces of δ cut out. Thus $\hat{\omega}$ is ω cut into pieces at all instances of δ, order preserved.

2. $\tilde{\omega} = \omega_1 \omega_2 \ldots \omega_n$ is ω with all instances of δ removed, a single string.

Definition: $\hat{L}(G) = \left\{ \hat{\omega} \mid \hat{\omega} \in 2^{V_T^*} \& \ S \overset{*}{\Rightarrow} \omega \right\}$

$$\tilde{L}(G) = \left\{ \tilde{\omega} \mid \tilde{\omega} \in V_T^* \& \ S \overset{*}{\Rightarrow} \omega \right\}$$

Context-Free Cut Grammars

An example of palindromic cut grammars follows:

$$G_{pal} = \left(\{S\}, \{a, \ b\}, \mathcal{P}_{pal}, S\right)$$

where: $\mathcal{P}_{pal} = \left\{S \ \Rightarrow \ aSa \mid bSb \mid \delta\right\}$

A sample derivation:

$$S \ \Rightarrow \ aSa \Rightarrow abSba \Rightarrow abaSaba \Rightarrow abaaSaaba \Rightarrow abaa \ \delta \ aaba \ \in L\left(G_{pal}\right)$$

In particular, $L\left(G_{pal}\right) = \left\{\omega\delta\omega^R\right\}$

Note that $\langle abaa, \ aaba\rangle \ \in \hat{L}\left(G_{pal}\right)$ and also that $abaa \cdot aaba \in \tilde{L}\left(G_{pal}\right)$.

Thus applying context-free cut grammars to DNA, we have:

$$G_{DNA}^1 = \left(\{S\}, \ \{a, \ c, \ g, \ t\} \cup \{\bar{a}, \bar{c}, \bar{g}, \bar{t}\}, \ \mathcal{P}_{DNA}^1, \ S\right)$$

where: $\mathcal{P}_{DNA}^1 = \left\{S \ \Rightarrow \ aS\bar{a} \mid cS\bar{c} \mid gS\bar{g} \mid tS\bar{t} \mid \delta\right\}$

However, using the Watson-Crick base complements:

$$\begin{aligned}
\bar{a} &= t \ , \ \bar{c} = g \\
\bar{t} &= a \ , \ \bar{g} = c
\end{aligned}$$

we get a nicer context-free cut grammar as follows

$$G_{DNA}^0 = \left(\{S\}, \ \{a, \ c, \ g, \ t\}, \ \mathcal{P}_{DNA}^0, \ S\right)$$

where: $\mathcal{P}_{DNA}^0 = \left\{S \ \Rightarrow \ aSt \mid cSg \mid gSc \mid tSa \mid \delta\right\}$

A sample derivation follows.

S   ⇒   aSt ⇒ atSat ⇒ attSaat ⇒ attgScaat ⇒ attgaStcaat
    ⇒   attga δ tcaat

Note that attga δ tcaat $\in L\left(G_{DNA}^{0}\right)$ and that $\left\langle attga,\ tcaat\right\rangle \in \hat{L}\left(G_{DNA}^{0}\right)$.

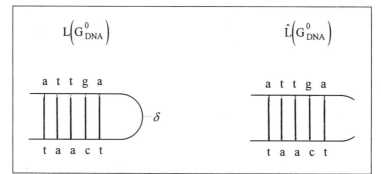

Figure 5.3 Cuts take place at the δ sites

Nick Grammars

DNA molecules have a "backbone" of deoxyribose saccharides linked with phosphates, just as RNA molecules have a "backbone" of ribose saccharides linked with phosphates. In both cases, the molecular structure is maintained by a number of forces in addition to the backbone, such as complementary base pair attractions, hydrogen bonding, etc. If the saccharide "backbone" is occasionally severed (bonds are broken), the molecule still maintains integrity, due to the other forces (complementary base pairs, hydrogen bonds, etc.). A DNA double helix, or RNA molecule that has such severed "backbone" bonds is referred to as a "nicked" DNA or a "nicked" RNA. Diagramatically:

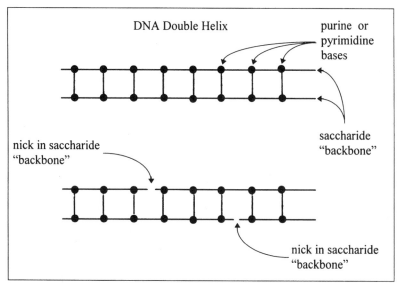

Figure 5.4 Nicks in double-helix structure

We can construct context-free cut grammars to model nicks [164, p. 32][a].

G^1_{nick} is a nick grammar for nicks in a single strand. Then

$$G^1_{nick} = \left(\{S\}, \{a, \ c, \ g, \ t\}, \mathcal{P}^1_{nick}, S \right),$$

where: $\mathcal{P}^1_{nick} = \left\{ S \ \Rightarrow \ aSt \ | \ cSg \ | \ gSc \ | \ tSa \ | \ S\delta \ | \ \delta \right\}$

A sample derivation:

$$S \ \Rightarrow \ aSt \Rightarrow acSgt \Rightarrow acS\delta gt \Rightarrow act\,S\,a\,\delta\,gt \Rightarrow act\,gSca\,\delta\,gt$$
$$\Rightarrow \ actg\,\delta\,ca\,\delta\,gt$$

thus (see Figure 5.5):

---

[a] Searls refers to this as G_{n1}.

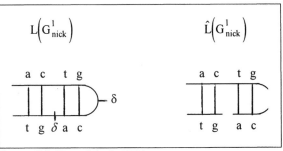

Figure 5.5 Two nicks in DNA

G_{nick}^2 is a nick grammar for nicks that can take place on both strands [164, p. 32],[b] then

$$G_{nick}^2 = \left(\{S\}, \left\{a, \ c, \ g, \ t\right\}, \ \mathcal{P}_{nick}^2, \ S\right),$$

$$\text{where: } \mathcal{P}_{nick}^2 = \left\{S \ \Rightarrow \ aSt \ | \ cSg \ | \ gSc \ | \ tSa \ | \ \delta S \ | \ S\delta \ | \ \delta\right\}$$

A sample derivation:

$$
\begin{aligned}
S \ &\Rightarrow \ aSt \Rightarrow a\,tSat \Rightarrow a\,t\,\delta\,S\,at \Rightarrow a\,t\,\delta\,cS\,gat \Rightarrow a\,t\,\delta\,ccS\,ggat \\
&\Rightarrow \ a\,t\,\delta\,cctSagg\,at \Rightarrow a\,t\,\delta\,cctS\,\delta\,aggat \Rightarrow a\,t\,\delta\,cctgSc\,\delta\,aggat \\
&\Rightarrow \ at\,\delta\,cctg\,\delta\,c\,\delta\,aggat
\end{aligned}
$$

Thus:

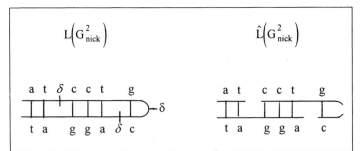

Figure 5.6 Nicks on either strand of a double-helix

Note that if nicks on both strands match up, then the double-stranded DNA breaks (see Figure 5.7).

---

[b] Searls refers to this as G_{n2}.

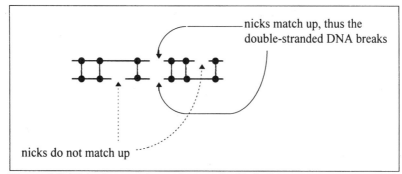

Figure 5.7 Nicks that match can cleave a double helix

Nicks that match up will cleave a DNA double helix, but a context-sensitive cut grammar that avoids this, and which has overhangs (sticky ends) of a minimum specified length is as follows [164, p. 32][c].

$$G_{\text{no break}} = \left(\{S\}, \ \{a, \ c, \ g, \ t\}, \ \mathcal{P}_{\text{no break}}, \ S \right),$$

where: $\mathcal{P}_{\text{no break}} = \left\{ S \Rightarrow aSt \mid cSg \mid gSc \mid tSa \mid \omega \delta S \overline{\omega}^{R} \mid \omega S \delta \overline{\omega}^{R} \mid \delta \right\}$

and $\omega \in V_{T}^{n}$, and n is the overhang length (minimal distance between nicks).

Example, with n = 3:

$$S \Rightarrow aSt \mid cSg \mid gSc \mid tSa \mid acc \delta S ggt \mid tatS \delta ata \mid \delta$$

a sample derivation:

$$
\begin{aligned}
S \Rightarrow{}& acc \delta S ggt \Rightarrow acc \delta tatS \delta ataggt \\
\Rightarrow{}& acc \delta tatacc \delta Sggt \delta ataggt \\
\Rightarrow{}& acc \delta tatacc \delta\delta ggt \delta ataggt
\end{aligned}
$$

See Figure 5.8.

---

[c] Searls refers to this as G_{n3}.

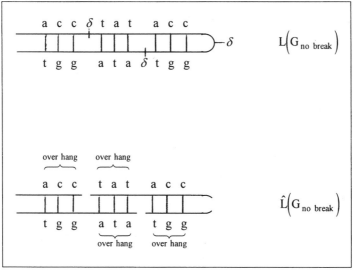

Figure 5.8 Language without cleavage due to matched nicks

Non-Linear Context-Free Cut Grammars

Hybrid double strands of molecules, one strand being DNA, the other strand being RNA are possible. In such cases, DNA substrands may not have matched RNA substrand complements. The extra substrands of DNA are called R loops. R loops were the first examples of introns [164, p. 33]. We have already encountered R loops or R rings, when we discussed Garzon's in vitro applications of FSA (see the previous chapter).

$$G_{hyb} = \left(\{S, R\}, \{a, c, g, t, u\}, \mathcal{P}_{hyb}, S \right),$$

$$\text{where:} \quad \mathcal{P}_{hyb} = \begin{cases} S & \Rightarrow & aSu \mid cSg \mid gSc \mid tSa \mid RS \mid \delta \\ R & \Rightarrow & Ra \mid Rc \mid Rg \mid Rt \mid \lambda \end{cases}$$

A sample derivation may be found in the following and in Figure 5.9.

S $\Rightarrow$ aSu $\Rightarrow$ atSau $\Rightarrow$ atcSgau $\Rightarrow$ atcRSgau $\Rightarrow$ atcRaSgau

$\Rightarrow$ atcRtaSgau $\Rightarrow$ atcRataSgau $\Rightarrow$ atcRtataSgau

$\Rightarrow$ atc λ tataSgau $\Rightarrow$ atctatagScgau $\Rightarrow$ atctataggSccgau

$\Rightarrow$ atctataggtSaccgau $\Rightarrow$ atctataggtcSgaccgau

$\Rightarrow$ atctataggtcaSugaccgau $\Rightarrow$ atctataggtca δ ugaccgau

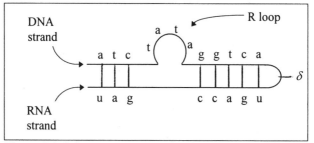

Figure 5.9 R loops are an example of non-linearity

Branched Hybridization Cut Languages

Other palindromic context-free cut grammars are possible.

$$G_Y = \Big(\{S, A\}, \ \{a, \ c, \ g, \ t\}, \ \mathcal{P}_Y, \ S \Big),$$

where: $\mathcal{P}_Y = \begin{cases} S & \Rightarrow \ aSt \mid cSg \mid gSc \mid tSa \mid AA \\ A & \Rightarrow \ aAt \mid cAg \mid gAc \mid tAa \mid \delta \end{cases}$

Branching structures such as the following are obtained.

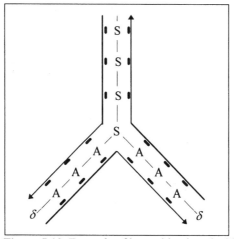

Figure 5.10 Example of brasnching in a double-helix

$$G_{\text{arb branch}} = \left(\{S\}, \ \{a, \ c, \ g, \ t\}, \ \mathcal{P}_{\text{arb branch}}, \ S \right),$$

where: $\mathcal{P}_{\text{arb branch}} = \left\{ S \ \Rightarrow \ aSt \mid cSg \mid gSc \mid tSa \mid SS \mid \delta \right\}$

Arbitrary branching structures are supported by this grammar, structures such as:

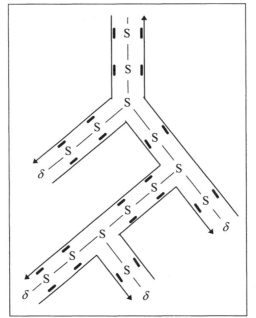

Figure 5.11 Arbitrary branching in a double-helix

$$G_{Fork} = \left(\{S\}, \; \{a, \; c, \; g, \; t\}, \; \mathcal{P}_{Fork}, \; S \right),$$

where: $\mathcal{P}_{Fork} = \{S \; \Rightarrow \; aSt \mid cSg \mid gSc \mid tSa \mid S\delta S \mid \lambda\}$

Note that the forks occur with $S\delta S$, and terminations with λ causes ends not to be cut.

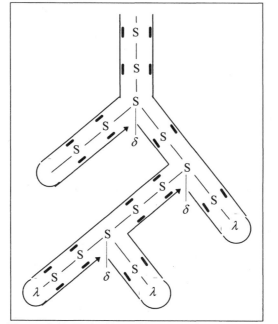

Figure 5.12 Forked branching where λ means no break

$$G_{arb} = \left(\{S\}, \{a, c, g, t\}, \mathcal{P}_{arb}, S \right),$$

where: $\mathcal{P}_{arb} = \{S \; \Rightarrow \; aSt \mid cSg \mid gSc \mid tSa \mid S\delta S \mid S\delta \mid \delta S \mid \lambda\}$

This grammar will insert cuts at arbitrary places.

Circular Context-Free Cut Languages

Definition: Given $\omega = \omega_1 \delta \omega_2 \delta \omega_3 \delta \ldots \delta \omega_n$ with $n > 1$, and $\omega_i \in V_T^*$

then $\breve{\omega} = \langle \omega_n, \omega_1, \omega_2, \ldots, \omega_{n-1} \rangle$ (circular cuts)

Example:

$$G_{Hol} = \left(\{S, A\}, \{a, c, g, t\}, \mathcal{P}_{Hol}, S \right),$$

where: $\mathcal{P}_{Hol} = \begin{cases} S & \Rightarrow & AS \mid \lambda \\ A & \Rightarrow & aAt \mid cAg \mid gAc \mid tAa \mid \delta \end{cases}$

Now for a derivation, and an examination of the corresponding structure (subscripts are to help clarify the structure).

$$
\begin{aligned}
S \Rightarrow{}& A_1 S_1 \Rightarrow A_1 A_2 S_2 \Rightarrow A_1 A_2 A_3 S_3 \Rightarrow A_1 A_2 A_3 A_4 S_4 \Rightarrow A_1 A_2 A_3 A_4 \; \lambda \\
={}& A_1 A_2 A_3 A_4 \Rightarrow tAaA_2 A_3 A_4 \Rightarrow taAtaA_2 A_3 A_4 \Rightarrow tatAataA_2 A_3 A_4 \\
\Rightarrow{}& tataAtataA_2 A_3 A_4 \Rightarrow tata\,\delta\,tata \cdot A_2 A_3 A_4 \Rightarrow tata\,\delta\,tata \cdot cAgA_3 A_4 \\
\Rightarrow{}& tata\,\delta\,tata \cdot ccAggA_3 A_4 \Rightarrow tata\,\delta\,tata \cdot ccgAcggA_3 A_4 \\
\Rightarrow{}& tata\,\delta\,tata \cdot ccggAccggA_3 A_4 \Rightarrow tata\,\delta\,tata \cdot ccgg\,\delta\,ccgg \cdot aAtA_4 \\
\Rightarrow{}& tata\,\delta\,tata \cdot ccgg\,\delta\,ccgg \cdot a\,c\,A\,g\,tA_4 \Rightarrow tata\,\delta\,tata \cdot ccgg\,\delta\,ccgg \cdot actAagtA_4 \\
\Rightarrow{}& tata\,\delta\,tata \cdot ccgg\,\delta\,ccgg \cdot actgAcagtA_4 \Rightarrow tata\,\delta\,tata \cdot ccgg\,\delta\,ccgg \cdot actg\,\delta\,cagt \cdot A_4 \\
\Rightarrow{}& tata\,\delta\,tata \cdot ccgg\,\delta\,ccgg \cdot actg\,\delta\,cagt \cdot aAt \\
\Rightarrow{}& tata\,\delta\,tata \cdot ccgg\,\delta\,ccgg \cdot actg\,\delta\,cagt \cdot acAgt \\
\Rightarrow{}& tata\,\delta\,tata \cdot ccgg\,\delta\,ccgg \cdot actg\,\delta\,cagt \cdot acgAcgt \\
\Rightarrow{}& tata\,\delta\,tata \cdot ccgg\,\delta\,ccgg \cdot actg\,\delta\,cagt \cdot acgtAacgt \\
\Rightarrow{}& tata\,\delta\,tata \cdot ccgg\,\delta\,ccgg \cdot actg\,\delta\,cagt \cdot acgtaAtacgt \\
\Rightarrow{}& tata\,\delta\,tata \cdot ccgg\,\delta\,ccgg \cdot actg\,\delta\,cagt \cdot acgta\,\delta\,tacgt
\end{aligned}
$$

Now, for the diagram of the structure, see Figure 5.13.

Holliday Structures

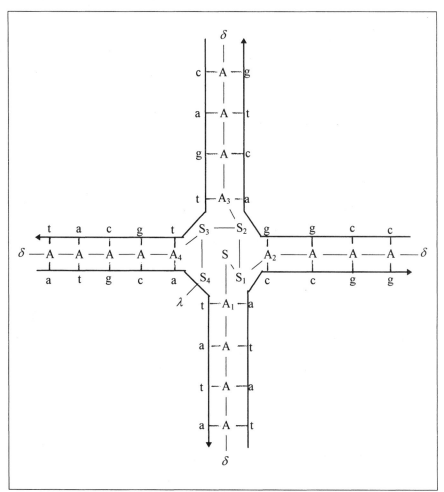

Figure 5.13 Holliday structure (circular cut grammar)

To be a true Holliday structure, both the "horizontal" and the "vertical" arms of the double strands must be identical or very close to identical. The context-free cut grammar provided is not adequate for the purpose. What is required is a context-sensitive cut grammar, which will be discussed in the next chapter.

Operons and Transformational Generative Grammars

A brief review of operons would be useful, then we shall examine how automata theory might be usefully employed in their study.

Escherichia coli growing in a glucose environment has little β-glactosease. However, grown in a lactose environment, then β-glactosease concentrations increase sharply. The presence of lactose induces the increase in β-glactosease. Indeed, β-glactosease is an inducible enzyme. In addition, galactoside permease is synthesized (it is required to transport lactose across the E. coli cell membrane). Also, thiogalactoside transacetylase is synthesized and this promotes the following acetylation reaction.

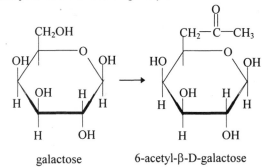

galactose   6-acetyl-β-D-galactose

In fact, there are three structural genes, referred to as z, y, and a.
The z gene is associated with β-glactosease protein.
The y gene is associated with glactoside permease protein.
The a gene is associated with the thiogalactoside transacetylase protein.

Exactly what does β-glactosease do? Lactose hydrolyzes into glucose and galactose.

β-galacosidase hydrolyzes lactose into galactose and lactose

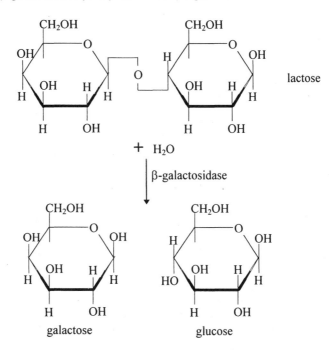

lactose

+ H₂O

β-galactosidase

galactose　　　　　glucose

The lac operon is composed of a number of genes, as seen below.
(See the discussion about operons in the first chapter.)

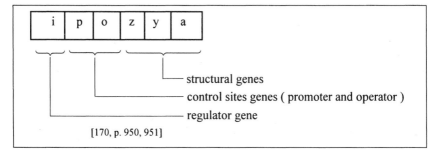

[170, p. 950, 951]

Figure 5.14 Genetic structure of the lac operon

E. coli mutants described as $z^- y^+ a^+$ lack β-galactose. However, E. coli "constitutive" mutants have also been found, which synthesize large amounts of all three proteins associated with the three structural genes. However, these mutants synthesize their three proteins with or without the presence of the lactose inducer. Jacob and Monod deduced that there existed a regulator "i" (inducer) gene. The normal E. coli is described as $i^+ z^+$

$y^+ a^+$, while the constitutive mutant is $i^- z^+ y^+ a^+$. Experiments demonstrated that the i^+ gene causes the synthesis of a repressor protein. Binding of the repressor to the operator prevents protein synthesis by the three constitutive structural z, y, a genes, while the promoter p gene is the RNA polymerase binding site. Inducer mRNA (i mRNA) protein is synthesized which binds to the repressor, thus repressor/operator binding is prevented, and as a result the three structural genes z, y, a are free to transcribe a polygenic mRNA (one protein for a number of genes). RNA polymerase is generally composed of parts: $\alpha\alpha\beta\beta' \sigma$, and we shall refer to the sigma component later.

In a series of papers [32-41], a language is constructed (using a transformational generative grammar) to describe operons (at least some operons). The grammar has changed in the papers; however, the idea is repeatedly expressed to use a transformational generative grammar ([33, pp. 211, 212] also [41, p. 404]). Linguists use generative transformational grammars with formatives in an attempt to describe speech. [d] Indeed, Collado-Vides wishes to use formatives too [32, p. 411]. Formatives (similar in certain respects to attribute grammars) support the positive (+) as opposed to negative (–) effects of various linguistic items. Thus Collado-Vides mentions +/– properties when describing chemical factors such as k_m, V_{max} [32, p. 411], and specifically when describing operon promoters Pr [35, p. 404], and RM [32, p. 411]. Thus Pr(+), Pr(–), RM(–), etc appear. The explicit use of analogies of speech that must be successive, non-overlapping ([33, pp. 211] and [41, p. 403]) and linear overlapping ([33, p. 223] and [41, p. 409]) might be troublesome insofar as these properties would seem to exclude non-human language systems as studied in splicing systems, graph splicing systems, Lindenmayer $\mathcal{L}$ systems, and much of syntactic pattern recognition. The reason why Collado-Vides wants attributes or formative-like properties will be explained soon. It is worthwhile pointing out that Collado-Vides feels that a finite state automaton cannot be used to describe operons, nor can context-free grammars be used but he does expect to use derivation trees [33, p. 214]. This can be explained by the use of a generative transformational grammar with formatives passing information between leaves of the derivational trees thereby limited, as well as with such notation as the following in production rules:

$$\{S_i\}_{n, \ n>0} \qquad\qquad [33, p. 216]$$

We should also note that ITRC means Initiation-of-Transcription Regulatory Category, and specifically, ITRC(+) = I, while ITRC(–) = Op [41, pp. 406, 407].

---

[d] In the subject of phonetics, a variety of linguistic descriptions referred to as "formatives" is needed to describe speech, such as tones (tonal glides or contours common to Asian languages, as opposed to tonal levels common to African languages), features such as moras, sandhi, registers (with up-stepping, downdrift, or terracing), etc. See *Tone: A Linguistic Survey*, by V. Fromkin, Academic Press, 1992, pp. 133-175, and *Vowels and Consonants*, by P. Ladefoged, Blackwell Publishers, 2000, p. 179.

lac operon DNA

Pr_r E_r . . . . I_{cap} Pr_{lac} Op_{lac} S_z S_y S_z

distance
(non−adjacency)

Figure 5.15 Detailed genetic regions in the lac operon

Legend [32, p. 405]

Pr   promoter gene (always present at the beginning of a transcription unit)
           Note: RNA polymerase must first bind to Pr before structural
               genes can transcribe into proteins.
Op   operator gene   (Note: Repressor binds to Op to stop transcription)
I   activator
S   structural gene

Subscripts
r   repressor gene
cap   catabolic activator protein
lac   lac operon

In addition, Collado-Vides uses the following to describe RNA order [33, p. 215]:

I_c initiation codon
T_c termination codon

In fact, Collado-Vides has proposed a variety of notations, as well as applications to a number of operons. Thus we might also find [33, pp. 217, 219]:

S_{trpE} when referring to the trp operon
B_{cap} when referring to binding regions
i   for repressor gene

$$G = \left(V_N, \ V_T, \ \mathcal{P}, \ O \right)$$

$$V_N = \left\{ O, \ T, \ R', \ S', \ R, \ Pr, \ Op, \ S, \ I \right\}$$

$$V_T = \left\{ I_{cap}, \ Pr_r, \ Pr_{lac}, \ Op_{lac}, \ S_r, \ S_z, \ S_y, \ S_a \right\}$$

where: $\mathcal{P}$ is as follows [32, p. 410].

$$
\begin{aligned}
O &\Rightarrow T\ T \\
T &\Rightarrow R'\ S' \\
R' &\Rightarrow I\ R \mid R \\
S' &\Rightarrow S\ S' \mid S \\
R &\Rightarrow Pr \mid Pr\ Op \\
Pr &\Rightarrow Pr_r\ Pr_{lac} \\
Op &\Rightarrow Op_{lac} \\
S &\Rightarrow S_r \mid S_z \mid S_y \mid S_a \\
I &\Rightarrow I_{cap}
\end{aligned}
$$

R' is used to refer to a pre-regulatory domain
R is a regulatory domain
S' is used to refer to a pre-structural region
S is a structural region

Thus we obtain derivation trees such as the following (slightly modified symbols, as different systems of symbols have been used in different papers).

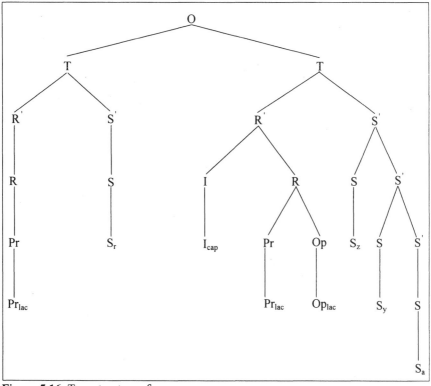

Figure 5.16 Tree structure of operon

Collado-Vides points out some explicit examples. Thus the Promoter Pr refers to:

araBAD E. coli	TTAGCGATCCTACCTGACGCTTTTTATCGCAACTCTCTACTGTTTCT
lac E. coli	TAGGCACCCCAGGCTTACACTTTATGCTTCCGGCTCGTATGTTGTG
λ phage	TAACACCGTGCGTGTTGACTATTTTACCTCTGGCGGTGATAATGGT

Figure 5.17 Operon promoter

Collado-Vides makes it very clear that distances are quite significant. Thus we find Figure 1 in [35, p. 410]:

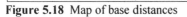

Figure 5.18 Map of base distances

Operon	Description of Operon	Origin
lac	I_{crp} Pr_{lac} Op_1 Op_2 S_z S_y S_a	E. coli
gal	Op_E Pr_2 $ITRC_{crp}$ Pr_1 Op_1 S_e S_t S_k	E. coli
glnA	I_{crp} Pr_1 NRI_1 NRI_2 Pr_2 S_{glnA} T_{glnA} Pr_3 Op_{glnG} S_{glnG} S_{glnL}	E. coli
proline	S_{putP} I_{crp} Pr_1 Op_{pro} Pr_2 S_{putA}	S. typhymurium
ara	S_{araC} Op_2 Op_1 Pr_c I_{crp} I_1 I_2 Pr_{bad} S_b S_a S_d	E. coli
Lambda	S_{ci} Op_{r3} Pr Pr_m Op_{r2} Op_{r1} S_{cro}	Lambda

Figure 5.19 Examples of some different operons

Thus a new grammar is proposed [35, p. 413].

$$G = \left(\left\{ O, \; X^{'}, \; S^{'} \right\}, \; \left\{ \left\{ I_{\tau} \right\}_{\tau=1}^{n}, \; \left\{ Op_{\tau} \right\}_{\tau=1}^{m}, \; Pr, \; \left\{ S_{\tau} \right\}_{\tau=1}^{j} \right\}, \; \mathcal{P}, \; O \right)$$

where: $\mathcal{P}$ is as follows.

$$O \; \Rightarrow \; X^{'} S^{'}$$
$$X^{'} \; \Rightarrow \; \left\{ I_{\tau} \right\}_{\tau=1}^{n} Pr \left\{ Op_{\tau} \right\}_{\tau=1}^{m} \; | \; \left\{ I_{\tau} \right\}_{\tau=1}^{n} Pr \; | \; Pr \left\{ Op_{\tau} \right\}_{\tau=1}^{m} \; | \; Pr$$
$$S^{'} \; \Rightarrow \; \left\{ S_{\tau} \right\}_{\tau=1}^{j}$$

Using the pumping theorem for context-free languages [34, pp. 322-324], R_i and T_i refer to genetic units of information with their associated proteins, and their targets, respectively. Collado-Vides shows that the pumping theorem for context-free grammars cannot apply. Specifically, if $uvxwy \in L$, then $uv^i xw^i y \in L$ for all i.

case 1: $v, w \in \{ R_{\tau} \}_{\tau}$; then there are no targets.

case 2: $v \in \{ R_{\tau} \}_{\tau}$, $w = \{ T_{\tau} \}_{\tau}$; then if $u x y \in L$ (allowing $\tau = 1, 2, 3$ for v, and $\tau = 1, 2$ for w), then $u \left\{ R \; R \; R \right\}^i x \left\{ T \; T \right\}^i y \notin L$; for example, let $i = 2$, we obtain:
$u R R R R R R x T T T T y$ and there are genetic units R without targets T.

case 3: let $v \in \{ T_{\tau} \}_{\tau}$, $w = \{ T_{\tau} \}_{\tau}$; then $u x y \in L$, but $u \{ T T T \}^i x \{ T T T \}^i y \notin L$. For example, if "i" is too large (a biochemical reason), this string can't be regulated. Note also [41, p. 417] that it is the view of Collado-Vides that context-free grammars are rejected (not for "linguistic" reasons) but due to reasons of biology.

In the same reference [34, pp. 322-324], specifically on p. 323, Collado-Vides gives us some examples of amino acid regulatory protein sequences R and their corresponding targets (T). In addition, some mutations are provided, and additional information that is not included here (see Figure 5.20).

	Regulator Protein	Target
Mnt repressor phage P22	DPHF = Asp·Pro·His·Phe	AGGTCCAC
Mnt−bs2	DPPF = Asp·Pro·Pro·Phe	AGATCCAC
repressor for phage 434	KRIQLG = Lys·Arg·Ile·Gln·Leu·Gly	ACAATAT
	KRIALG = Lys·Arg·Ile·Ala·Leu·Gly	TCAATAT
lac repressor (single mutant)	YQTVSRVV = Tyr·Gln·Thr·Val·Ser·Arg·Val·Val	TGTGAGC
	QMTVSRVV = Gln·Met·Thr·Val·Ser·Arg·Val·Val	TGTTAGC
	VATVSRVV = Val·Ala·Thr·Val·Ser·Arg·Val·Val	TGTAAGC
lac repressor (double mutant)	YQTVSNVV = Tyr·Gln·Thr·Val·Ser·Asn·Val·Val	TTTGAGC
	VATVSNVV = Val·Ala·Thr·Val·Ser·Asn·Val·Val	TTTTAGC

Figure 5.20 Examples of some regulator proteins

Continuing our study, we attempt to create a more accurate grammar to describe operons than the grammars constructed so far. A rough classification of regulatory units is made as to the location of these units. Regulatory units may be either proximal or remote. We need a little greater precision, thus a working definition of these terms.

Regulatory units are proximal if their position enables direct contact of the regulatory protein with RNA polymerase (the RNA transcription start site is at +1). Thus a proximal position is at about −65 to +20. A regulatory unit is remote if its position is not proximal.

Using the definition of proximal vs. remote, we obtain a revised grammar, with "features" or descriptors. An example of a derivation [36]:

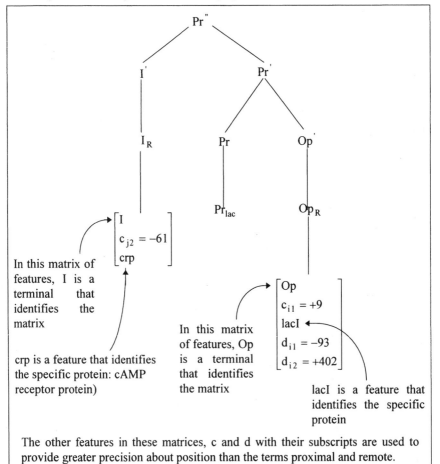

The other features in these matrices, c and d with their subscripts are used to provide greater precision about position than the terms proximal and remote.

Figure 5.21 Transformational grammar with features, for operons

The feature c_{in} is used to specify a proximal site, where the "c" specifies that the site is proximal.

The feature d_{in} is used to specify a distant (remote) site, where the "d" specifies that the site is distant.

Features that specify sites may be remote (r), proximal (p), or referential (R), but Collado-Vides decided to eliminate these descriptors later [37, p. 98].

Two models or grammars are proposed. In model A, the distance has its own matrix of features, with "D" (distance) used to identify the matrix. This model is found to be unacceptable (although "D" type matrices are retained), and instead, model B is used, where the distances are specified as features, using "c" or "d" (see the previous derivation). To be explicit, see Figures 5.22 and 5.23. Note especially that the matrices in Figure 5.22 are expanded, to support coordinate "c" and distant "d" features.

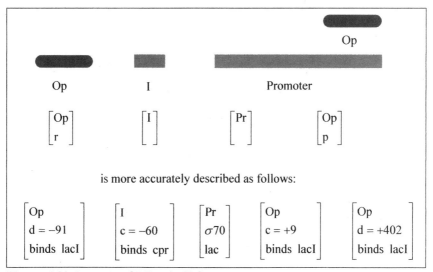

Figure 5.22 Presence of "c" and "d" features

Note that here "d" stands for distance and "c" stands for coordinate. However, it is decided that the feature "binds X" will be restricted to coordinates of the proximal site. In addition, as there are situations in which there is more than one regulator, an index will be used to show homology. Thus if there are two regulators, one will be identified with index "i" , the second by index "j". Thus there can be c_i, d_i, c_j, d_j features, as seen I Figure 5.23.

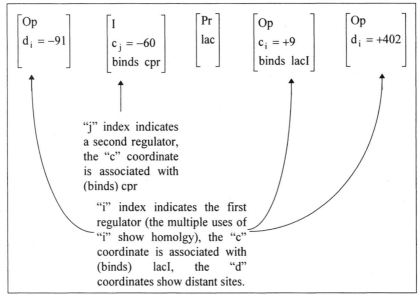

Figure 5.23 Grammar with multiple features

Thus a few representative descriptions are now provided [38].

Regulato	Promoter	Description (note: "ph" indicates phase - the number of helix turns that separate homologous sites from their reference)
AraC, CRP	araBAD	$\begin{bmatrix} Op \\ d_i = -210 \\ ph = 19.0 \end{bmatrix} \begin{bmatrix} I \\ d_i = -29 \\ crp \end{bmatrix} \begin{bmatrix} D \\ c_i = -64 \\ araC \end{bmatrix} \begin{bmatrix} I \\ c_i = -43 \\ \end{bmatrix} \begin{bmatrix} Pr \\ \\ \end{bmatrix}$
AraC, CRP	araC	$\begin{bmatrix} Op \\ d_i = -72 \\ \end{bmatrix} \begin{bmatrix} I \\ c_i = -70 \\ crp \end{bmatrix} \begin{bmatrix} Op \\ d_i = -21 \\ \end{bmatrix} \begin{bmatrix} Op \\ c_i = -29 \\ araC \end{bmatrix} \begin{bmatrix} Pr \\ \\ \end{bmatrix} \begin{matrix} Op \\ d_i = +136 \\ ph = 13.0 \end{matrix}$
DeoR CytR, CRP	deo p2	$\begin{bmatrix} Op \\ d_i = -879 \\ ph = 83.6 \end{bmatrix} \begin{bmatrix} Op \\ d_i = -300 \\ ph = 26.5 \end{bmatrix} \begin{bmatrix} Op \\ d_j = -52.5 \\ \end{bmatrix} \begin{bmatrix} Op \\ c_k = -65 \\ CytR \end{bmatrix} \begin{bmatrix} I \\ c_j = -41.5 \\ crp \end{bmatrix} \begin{bmatrix} Pr \\ \\ \end{bmatrix}$
MalT, CRP	malE	$\begin{bmatrix} \frac{1}{2}\ I \\ d_i = -189 \end{bmatrix} \begin{bmatrix} I \\ d_i = -160 \\ ph = 15.3 \end{bmatrix} \begin{bmatrix} I \\ d_j = -95 \\ ph = 10.8 \end{bmatrix} \begin{bmatrix} I \\ d_j = -61 \\ ph = 2.8 \end{bmatrix} \begin{bmatrix} I \\ d_j = -32 \\ crp \end{bmatrix} \begin{bmatrix} I \\ c_i = -44 \\ malT \end{bmatrix} \begin{bmatrix} Pr \\ \\ \end{bmatrix}$
MalT, CRP	malK	$\begin{bmatrix} I \\ d_i = -160 \\ ph = 15.3 \end{bmatrix} \begin{bmatrix} I \\ d_j = -128 \\ ph = 10.8 \end{bmatrix} \begin{bmatrix} I \\ d_j = -99 \\ ph = 2.8 \end{bmatrix} \begin{bmatrix} I \\ d_j = -6.5 \\ crp \end{bmatrix} \begin{bmatrix} I \\ c_i = -68 \\ malT \end{bmatrix} \begin{bmatrix} \frac{1}{2}\ I \\ c_i = -37.5 \end{bmatrix} \begin{bmatrix} Pr \\ \\ \end{bmatrix}$
Tyr	aroL	$\begin{bmatrix} Pr \\ \\ \end{bmatrix} \begin{bmatrix} Op \\ c_i = -4.5 \\ ph = 5.4 \end{bmatrix} \begin{bmatrix} Op \\ c_i = +12 \\ ph = 2.0 \end{bmatrix} \begin{bmatrix} Op \\ c_i = +31 \\ tyrR \end{bmatrix}$

Figure 5.24 Various combinations of features

Thus the grammar used by Collado-Vides to describe operons places great importance upon the position and relative distance of genes and relevant proteins. However, the reason for this has not yet been explored sufficiently. σ70 and σ54 bacterial operon promoters are studied, and an important property is found: "... the ability to be activated from protein bonding at remote sites ..." [39, p. 351]. Thus for the large σ70 collection of operons, all promoters require a site to bind to the regulator or close enough to the promoter to enable direct contact between the regulator and the polymerase. These proximal sites differ from remote sites. σ54 promoters are activated by remote sites. "The paradigm to explain this interaction is the idea that once the activator is bound to the remote site, the DNA in-between the activator and the RNA polymerase will bend, enabling the two proteins to interact directly." [39, p. 352] A number of suggestive mechanisms are proposed; see Figures 5.25 through 5.27. Note that in Figure 5.27, the intention is to show how the folding is thought to take place, allowing distant components to fold into close proximity, thereby allowing activation.

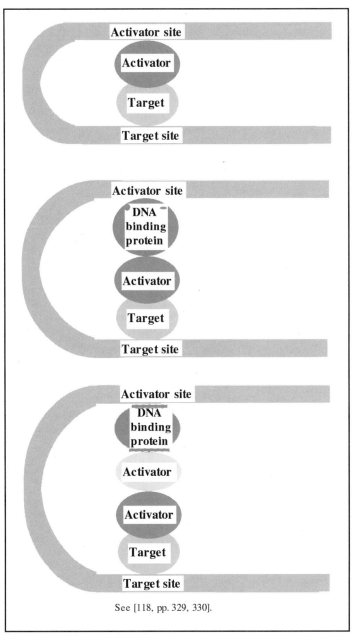

See [118, pp. 329, 330].

Figure 5.25 How bending allows distant features to come into close proximity

Wolffe suggests the following mechanism.

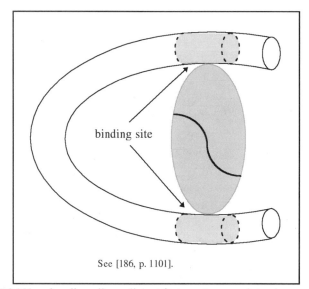

binding site

See [186, p. 1101].

Figure 5.26 How bending allows distant features to come into close proximity

Collado-Vides also agrees with this mechanism [39, p. 354] (see the following), and feels that a grammar to deal with operons must use features with coordinates and distances to reflect these important matters. We end our discussion about operon grammars at this point.

Collado-Vides suggests the following mechanism.

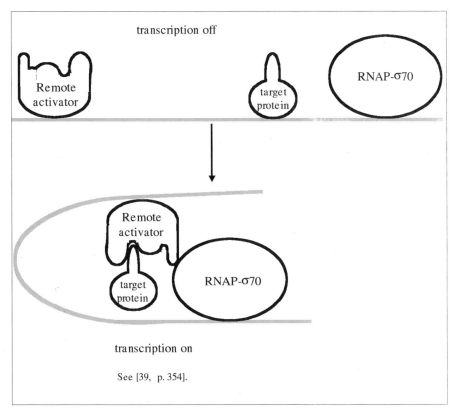

Figure 5.27 How bending allows distant features to come into close proximity

Cytochrome-C Context-Free Protein Grammar

Cytochrome-C is important for different reasons, as well as the fact that it is used to detoxify foreign substances [170, p. 704]. We are interested in applications of computation to different life forms.

The following context-free grammar is used to describe the protein sequence of human cytochrome-C [70, p. 61].

$$G = \left(\{S, \ \sigma_1, \ \sigma_2, \ \sigma_3 \}, \ V_T, \ \mathcal{P}, \ S \right)$$

where:

V_T = { amino acids } = { A, C, D, E, F, G, H, I , K, L, M, N, P, Q, R, S, T, V, W, Y }

and the production rules $\mathcal{P}$:

$S \ \Rightarrow \ $ GD$\sigma_1\sigma_3$IFIMKCSQCHTσ_1 GKH σ_2PNLHGLFGRσ_2QAPGYSYTAANKN
KGIIWGEDTLMEYLENPσ_3YIPGTKMIFVGIσ_3KEERADLIAYLσ_3ATNE

and

$\sigma_1 \ \Rightarrow \ $ VEKG
$\sigma_2 \ \Rightarrow \ $ KTG
$\sigma_3 \ \Rightarrow \ $ KK

Similarly, the Escherichia coli gene coding for the DNA antecedent of tyrosine t-RNA may be viewed as a context-free grammar [70, p. 63].

$$G = \left(\{S\} \cup \{\sigma_i\}_{i=1}^{5}, \ \{A, \ C, \ G, \ T\}, \ \mathcal{P}, \ S \right)$$

where the production rules $\mathcal{P}$ are as follows.

$$S \ \Rightarrow \ \sigma_5 \sigma_5 \sigma_5 \sigma_{10} \ G \ \sigma_9 \sigma_{10} \ A \ \sigma_3 \sigma_8 \sigma_3 \ G \ \sigma_6 \ A \ \sigma_{11} \ AC \ \sigma_{10} \ CAGA \ \sigma_4 \ GAG$$
$$\sigma_7 \sigma_1 \sigma_5 \sigma_8 \ GC \ \sigma_2 \sigma_8 \sigma_8 \ A \ \sigma_8 \ AC \ \sigma_{10} \ GT \ \sigma_9 \sigma_{11} \ CT \ \sigma_4 \ C \ \sigma_5 \sigma_8 \sigma_1 \ A \ \sigma_2 \ GC$$

and

$$\sigma_1 \ \Rightarrow \ \sigma_1 \ C \ \sigma_7 \sigma_8 \ TT$$
$$\sigma_2 \ \Rightarrow \ \sigma_6 \sigma_{10} \sigma_9$$
$$\sigma_3 \ \Rightarrow \ T \sigma_6 \sigma_9$$
$$\sigma_4 \ \Rightarrow \ TTTA$$
$$\sigma_5 \ \Rightarrow \ \sigma_{11} \ G$$
$$\sigma_6 \ \Rightarrow \ \sigma_7 \ G$$
$$\sigma_7 \ \Rightarrow \ TC$$
$$\sigma_8 \ \Rightarrow \ CC$$
$$\sigma_9 \ \Rightarrow \ AA$$
$$\sigma_{10} \ \Rightarrow \ GG$$
$$\sigma_{11} \ \Rightarrow \ TG$$

The last example is for human ribosomal RNA 55 rRNA KB carcinoma [70, p. 64].

The grammar is also context-free, as follows:

$$G = \left(\{S\} \cup \{\sigma_i\}_{i=1}^{12}, \ \{A, \ C, \ G, \ U\}, \ \mathcal{P}, \ S \right)$$

where the production rules $\mathcal{P}$ are as follows:

$$
\begin{aligned}
S \Rightarrow \ & \sigma_9\sigma_2\sigma_{11} \, G\sigma_5\sigma_{10}\sigma_3\sigma_3\sigma_4 \, A\sigma_{11}\sigma_7\sigma_5 \, C\sigma_8\sigma_9\sigma_2\sigma_8\sigma_9\sigma_2\sigma_8\sigma_6 \\
& A\sigma_7\sigma_{12} \, A\sigma_7 \, A\sigma_1\sigma_9 \, C\sigma_1\sigma_5 \, UG\sigma_9\sigma_{12}\sigma_9 \, A\sigma_2 \, U\sigma_6 \, UG \\
& \sigma_6 \, G\sigma_{11} \, C\sigma_7\sigma_4\sigma_6\sigma_{10}\sigma_3\sigma_1\sigma_9 \, G\sigma_4\sigma_{12}\sigma_1\sigma_2 \, U
\end{aligned}
$$

and

$$
\begin{aligned}
\sigma_1 &\Rightarrow GG \\
\sigma_2 &\Rightarrow CU \\
\sigma_3 &\Rightarrow \sigma_{11} \, C \\
\sigma_4 &\Rightarrow \sigma_2 \, G \\
\sigma_5 &\Rightarrow \sigma_7 \, C \\
\sigma_6 &\Rightarrow \sigma_1 \, A \\
\sigma_7 &\Rightarrow GC \\
\sigma_8 &\Rightarrow G \, \sigma_{10}\sigma_2 \, C \\
\sigma_9 &\Rightarrow GU \\
\sigma_{10} &\Rightarrow AU \\
\sigma_{11} &\Rightarrow AC \\
\sigma_{12} &\Rightarrow UA
\end{aligned}
$$

Pawlak's Method

Pawlak's devised a method to obtain the 20 amino acid encodings from 64 codons, [103, pp. 91-101].

i, j, k ∈ { 0, 1, 2, 3 }, with i < j, snd j ≥ k

Thus,	reordering:	j	i	k	Symbol
i = 0, j = 1, k = 0, 1		1	0	0	a
j = 2, k = 0, 1, 2		1	0	1	b
j = 3, k = 0, 1, 2, 3		2	0	0	c
		2	0	1	d
i = 1, j = 2, k = 0, 1, 2		2	0	2	e
j = 3, k = 0, 1, 2, 3		2	1	0	f
		2	1	1	g
i = 2, j = 3, k = 0, 1, 2, 3		2	1	2	h
		3	0	0	i
		3	0	1	j
		3	0	2	k
		3	0	3	l
		3	1	0	m
		3	1	1	n
		3	1	2	o
		3	1	3	p
		3	2	0	q
		3	2	1	r
		3	2	2	s
		3	2	3	t

Of course, Pawlak's symbols a, b, c, d, e, f, g, h, i, j, k, l, m, n, o, p, q, r, s, t represent the 20 amino acids (see Chapter 1). The twenty collected amino acid representations may be seen in Figure 5.28. One might ask, what about additional amino acids, such as the 21st and 22nd, mentioned in Chapter 3 (Figure 3.19)? Specifically, it must be possible to expand the possible amino acids to take account of other amino acids which, although rare, are found to exist naturally in nature, in addition to artificially created proteins with new amino acids that may be created by mechanisms discussed in Chapter 3.

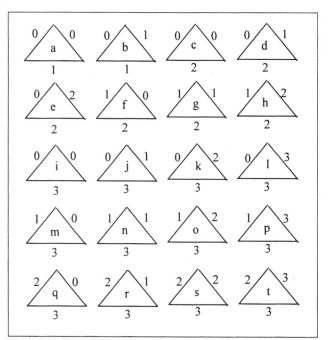

Figure 5.28 Pawlak's Twenty Amino Acid Representations

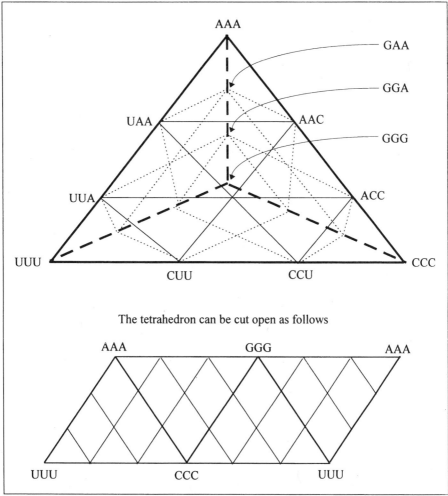

Figure 5.29 Base triples on a tetrahedron

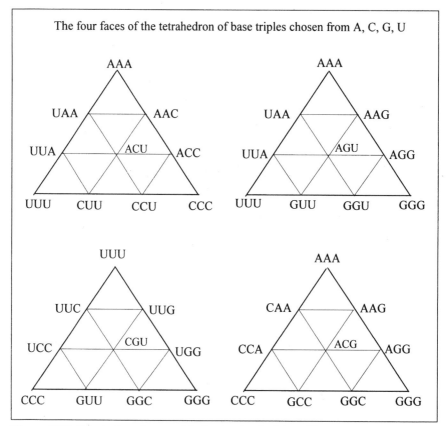

Figure 5.30 Tetrahedron faces and their associated triples of bases

Abbreviations of Permutations (plus the other 4)

UAA = UAA, AUA, AAU	AAG = AAG, GAA, AGA	AAA, CCC,
UUA = UUA, AUU, UAU	AGG = AGG, GAG, GGA	GGG, UUU
AAC = AAC, CAA, ACA	GGU = GGU, UGG, GUG	
ACC = ACC, CAC, CCA	GUU = GUU, UGU, UUG	
CCU = CCU, UCC, CUC	GGC = GGC, CGG, GCG	
CUU = CUU, UCU, UUC	GCC = GCC, CGC, CCG	

ACU = ACU, AUC, CAU, CUA, UAC, UCA
AGU = AGU, AUG, GAU, GUA, UAG, UGA
CGU = CGU, CUG, GCU, GUC, UCG, UGC
ACG = ACG, AGC, CAG, CGA, GAC, GCA

Pawlak provides a recursive definition of well-formed strings using the 20 triangles labeled "a" through "t". These strings of triangles constitute a protein language, found to be equivalent to a propagating, nondeterministic semi-Lindenmayer system (P0L). Pawlak's recursive definition follows.

1. All of the 20 triangles above are well-formed strings.
2. Given two well-formed strings "x" and "y", then the numbers of the strings of triangles   must match up. Some examples will make this clear.
3. Only those strings (of amino acids) obtained by the above two rules are well formed.

Examples:

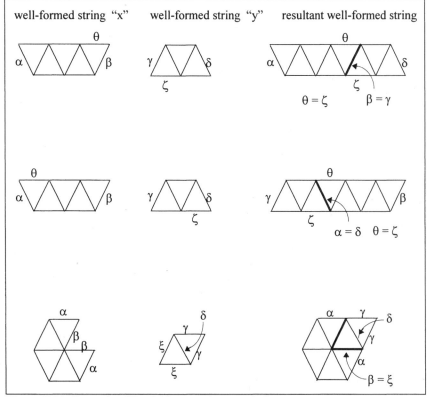

Figure 5.31 Example of Pawlak's recursive protein language

Rather than continue with Pawlak's ideas, we shall pick up with a variant view, that of Vauquois.

Vauquois Context-Free Grammar

Vauquois has a Chomsky context-free grammar for proteins, as follows:

G = ({S, A, B, C}, {a, b, c, d, e, f, g, h, i, j, k, l, m, n, o, p, q, r, s, t}, $\mathcal{P}$, S)

where: $\mathcal{P}$ is as follows.

S ⇒ a | c | i | l | dA | fA | jA | mA | eB | kB | qB | lC | gAA | nAA | hAB | hBA |
    oAB | oBA | pAC | pCA | rAB | rBA | sBB | tBC | tCB

A ⇒ a | bA

B ⇒ c | dA | fA | eB | gAA | hAB | hBA

C ⇒ i | jA | mA | kB | qB | lC | nAA | oAB | oBA | pAC | pCA | rAB | rBA | sBB | tBC | tCB

The following is a sample derivation. Any one-to-one mapping between Pawlak's symbols for amino acids and actual amino acids will be a derivation of a polypeptide

S ⇒ tBC   ⇒ thBAC   ⇒ theBAC

  ⇒ thegAAAC   ⇒ thegbaAAAC   ⇒ thegbaAAC

  ⇒ thegbabAAC   ⇒ thegbabaAC   ⇒ thegbabaaC

  ⇒ thegbabaapAC   ⇒ thegbabaapaC   ⇒ thegbabaapalC

  ⇒ thegbabaapali

It has been noted that there are shortcomings for both the views of Pawlak as well as Vauquois. The shortcomings will now be quickly reviewed.

Although we shall not examine non-automata-theoretic attempts to deal with complicated three-dimensional protein conformations, there are papers that can be consulted [93].

A grammar has been partially specified to show how proteins may be encoded by a gene. This partially specified context-sensitive grammar follows (promoters are omitted; see discussion of operons [160, p. 585]).

<gene>   ⟹ <upstream> <transcript> <downstream>
<transcript>   ⟹ <5'-untranslated-region> <start-codon> <coding-region> <3'-untranslated-region>
<start-codon>   ⟹ met
<coding-region>   ⟹ <codon> <coding-region> | <stop-codon> | <splice> <coding-region>
<codon>   ⟹ <lys> | <asn> | <ile> | <thr> | <met> | <ser> | <gln> | <his> | <arg> | <pro> |
                      <asp> | <glu> | <ala> | <gly> | <val> | <tyr> | <trp> | <cys> | <phe> | <leu>

<stop-codon>   ⟹ taa | tag | tga
<splice>   ⟹ <intron>
<intron>   ⟹ gt <intron-body> ag
<lys>   ⟹ aa <purine>   <asn> ⟹ aa <pyrimidine>
<ile>   ⟹ at <pyrimidine> | ata   <thr> ⟹ ac <base>
<met>   ⟹ atg   <ser> ⟹ ag <pyrimidine> | tc <base>
<gln>   ⟹ ca <purine>   <his> ⟹ ca <pyrimidine>
<arg>   ⟹ cg <base> | ag <purine>   <pro> ⟹ cc <base>
<asp>   ⟹ ga <pyrimidine>   <glu> ⟹ ga <purine>
<ala>   ⟹ gc <base>   <gly> ⟹ gg <base>
<val>   ⟹ gt <base>   <tyr> ⟹ ta <pyrimidine>
<trp>   ⟹ tgg   <cys> ⟹ tg <pyrimidine>
<phe>   ⟹ tt <pyrimidine>   <leu> ⟹ tt <purine> | ct <base>
<base> ⟹ <purine> | <pyrimidine>

<purine>   ⟹ a | g   <pyrimidine> ⟹ c | t
<splice> a ⟹ a <intron>   <splice> c ⟹ c <intron>
<splice> g ⟹ g <intron>   <splice> t ⟹ t <intron>
a <splice> ⟹ <intron> a   c <splice> ⟹ <intron> c
g <splice> ⟹ <intron> g   t <splice> ⟹ <intron> t

Amino Acid Sequences for Cytochrome-C

Three considerations ought to be taken into account during the synthesis of proteins.

1. The nucleotide bases A, C, G, and T form Watson-Crick complements, as do the RNA bases A, C, G, and U. Thus:

A may be replaced by U, U by A
C may be replaced by G, G by C

If such a systematic replacement by complements is done, then the resultant (complementary) triangles are as well formed as their originals. This was not taken into consideration.

2. The triangle of nucleotide bases can be weighted–but it wasn't.

3. Protein formation need not proceed serially (in one direction), rather a simultaneous formation (as in Lindenmayer systems and shape grammars) can take place. There should be a closer parallel with dynamic steric conformation.

As a last note of interest, there is a topic that has not been emphasized as little work of a computational-theoretic nature exists. However, it is quite possible that a grammar approach might be developed in the future, and thus a small note might not be inappropriate.

In the first non-review chapter of this book (Chapter 3), the possibility of early life evolution on this planet (as well as extraterrestrial life forms) based upon threose and hexose helices was mentioned. TNA analogues of DNA were mentioned. Evolution of life may be investigated by examining the molecules of living things to determine similarities, and thereby creating a phylogenetic tree. Such a view has been taken, examining amino acid polypeptide sequences.

Collecting sequences of amino acids for comparison, specifically, cytochrome-c proteins for 21 life forms, an examination finds interesting facts, supporting the creation of a phylogenetic tree [45]. The probabilities that cytochrome-c would have evolved independently in these different life forms is exceedingly remote. The amino acid sequences compared exceed 100 amino acids; thus the number of different possible chains is large: 20^{100}. It is an almost certainty that these life forms evolved from a common ancestor. The existence of a tree of life might provide the basis for a context-free grammar to describe the systematics. In any case, some interesting observations:

common amino acids at position 15
vertebrates and insects	K (lysine)
fungi and wheat	A (alanine)

common amino acids at position 17
fungi, yeast, Candida	L (leucine)
wheat and most animals	I (isoleucine)

common amino acids at position 93
insects and plants	L (leucine)
vertebrates	I (isoleucine)

See the following.

Based upon cytochrome-c, the following taxonomy results:

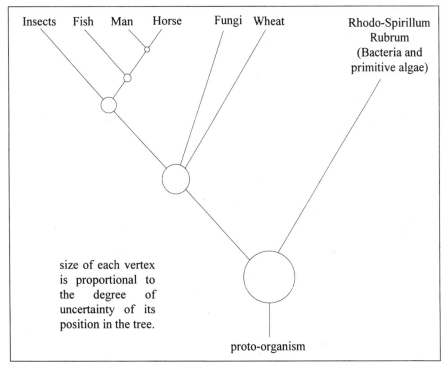

Figure 5.32 Evolutionary tree metric based upon cytochrome-c using PAMs

The unit of branch length is measured in accepted point mutations per 100 amino acid positions (PAMs). This length has been roughly correlated with time, assuming that the rate of mutation for the comparatively stable cytochrome-c remains constant across different species. It has been estimated that 11.5 PAM corresponds to 400 million years.

Amino Acid Position:

```
                       3           10        20        30        40        50        60
                       3 4 5 6 7 8 9 0 1 2 3 4 5 6 7 8 9 0 1 2 3 4 5 6 7 8 9 0 1 2 3 4 5 6 7 8 9 0 1 2 3 4 5 6 7 8 9 0 1 2 3 4 5 6 7 8 9 0
```

Species	Sequence (residues 3–60)
Human	F S E A P P G D V E K G K K I F I M K C S Q C H T V E K G G K H K T G P N L H G L F G R K T G Q A P G Y S Y T A A N
Rhesus Monkey	F S E A P P G D V E K G K K I F I M K C S Q C H T V E K G G K H K T G P N L H G L F G R K T G Q A P G Y S Y T A A N
Horse	F S E A P P G D V E K G K K I F V Q K C A Q C H T V E K G G K H K T G P N L H G L F G R K T G Q A P G F T Y T D A N
Pig, Bovine, Sheep	F S E A P P G D V E K G K K I F V Q K C A Q C H T V E K G G K H K T G P N L H G L F G R K T G Q A P G F S Y T D A N
Dog	F S E A P P G D V E K G K K I F V Q K C A Q C H T V E K G G K H K T G P N L H G L F G R K T G Q A P G F S Y T D A N
Gray Whale	F S E A P P G D V E K G K K I F V Q K C A Q C H T V E K G G K H K T G P N L H G L F G R K T G Q A V G F S Y T D A N
Rabbit	F S E A P P G D V E K G K K I F V Q K C A Q C H T V E K G G K H K T G P N L H G L F G R K T G Q A P G F S Y T D A N
Kangaroo	F S E A P P G D I E K G K K I F V Q K C A Q C H T V E K G G K H K T G P N L H G L F G R K T G Q A P G F T Y T D A N
Chicken, Turkey	F S E A P P G D I E K G K K I F V Q K C S Q C H T V E K G G K H K T G P N L H G L F G R K T G Q A E G F S Y T D A N
Penguin	F S E A P P G D I E K G K K I F V Q K C S Q C H T V E K G G K H K T G P N L H G L F G R K T G Q A E G F S Y T D A N
Peking Duck	F S E A P P G D V E K G K K I F V Q K C A Q C H T V E K G G K H K T G P N L H G L F G R K T G Q A E G F S Y T D A N
Snapping Turtle	F S E A P P G D V E K G K K I F V Q K C A Q C H T V E K G G K H K V G P N L H G I I G R K T G Q A E G F S Y T E A N
Bullfrog	F S E A P P G D V E K G K K I F V Q K C A Q C H T V E K G G K H K V G P N L Y G L I G R K T G Q A A G F S Y T D A N
Tuna	F S E A P P G D V A K G K K I F V Q K C A Q C H T V E N G G K H K V G P N L W G L F G R K T G Q A E G Y S Y T D A N
Screwworm fly	F S G V P A G D V E K G K K I F V Q R C A Q C H T V E A G G K H K V G P N L H G L F G R K T G Q A A G F A Y T N A N
Silkworm moth	F S G V P A G N A E N G K K I F V Q R C A Q C H T V E A G G K H K V G P N L H G F Y G R H S G Q A Q G Y S Y T D A N
Wheat	F S E A P P G N P D A G A K I F K T K C A Q C H T V D A G A G H K Q G P N L H G L F G R Q S G T T A G Y S Y S A A N
Fungus (Neurospora)	F S G F S A G D S K K G A N L F K T R C A E C H G E G G N L T Q K V G P A L H G L F G R K T G S V D G Y A Y T D A N
Fungus (Yeast)	F T E F K A G S A K K G A T L F K T R C E L C H T V E K G G P H K V G P N L H G I F G R H S G Q A E G Y S Y T D A N
Fungus (Candida)	P A P F E Q G S A K K G A T L F K T R C A E C H T I E A G G P H K V G P N L H G I F S R H S G Q A Q G Y S Y T D A N
Rhodospirillum	E G D D A A A G E K V S K – K K C L A C H T F D Q G A N K H K D D Y A Y S E S Y

For all of the above except the bacterium Rhodospirillum, the first two amino acids are alanine and serine. I provide only the first 60 amino acids.

Chapter 6: Context-Sensitive Languages: DNA, RNA, Proteins

Plague

"The spread of plague [231, pp. 128, 129] was accordingly explained as a result of corruption or infection of the air that altered for the worse the complexion of those who breathed it; the precipitating cause of the bad air was often, but not always, said to be astrological. In astrological theory, the precipitating cause of outbreaks of epidemic disease was usually held to be adverse conjunction of the planets; various medical and other writers produced tracts attributing the outbreak of the Black Death, which arrived in Sicily and southern Italy in 1347 and swept across different parts of Europe until 1351, to the conjunction of the three superior planets said to have occurred in 1345. Variations in individual complexion or horoscope were called on to explain why, when a whole community breathed the same air, some people got sick and others did not. This type of explanation was not as much at variance as is sometimes supposed with the belief in contagion from person to person, and from infected goods, held by those outside the medical profession and implicit in the quarantine regulations that began to be imposed by some public authorities before the end of the fourteenth century. Physicians recognized clearly that proximity to plague victims made one liable to get plague–hence the precautions they took when visiting the sick–and some of them termed it *contagium*, meaning that it passed rapidly from one person to another."

Secondary Structure Relationships

The secondary structure of bases induces constraints upon any linguistic characterization used to describe DNA, RNA, and proteins. Some of these constraints will now be analyzed.

String α' is related to string α, in that $\alpha' = \alpha^R$ (reversed order). In simple terms, the relationship between α and α' is palindromic; thus we see that as palindromes are supported by context-free languages, then DNA and RNA cannot in general be described by regular expressions. "Inverted repeats" illustrates this.

For example, if $\alpha = a\ c\ g\ t$, then $\alpha' = (a\ c\ g\ t)^R = t\ g\ c\ a = \bar{a}\ \bar{c}\ \bar{g}\ \bar{t}$.

Inverted Repeats

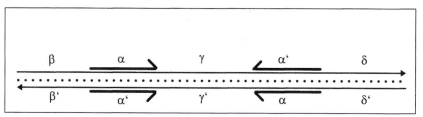

Figure 6.1 Inverted repeat primary structure

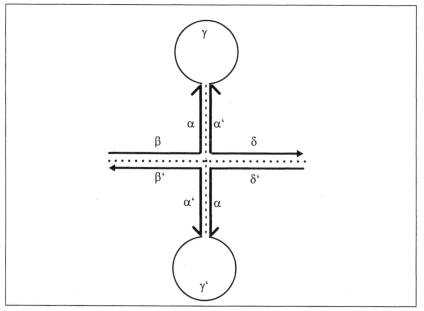

Figure 6.2 Inverted repeat secondary structure

Alternatively, inverted repeats may be viewed as follows [158, p. 71].

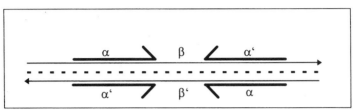

Figure 6.3 Inverted repeat primary structure

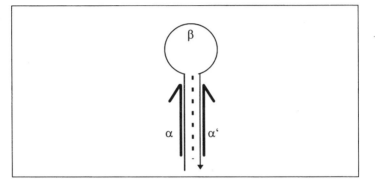

Figure 6.4 Inverted repeat secondary structure

Direct Repeats

Direct repeats are the occurrence of a substring α multiple times, on the same strand of the double-stranded DNA (or RNA).

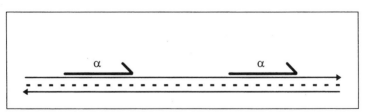

Figure 6.5 Direct repeat primary structure

Tandem Repeats

Tandem repeats refer to a substring α that occurs, when two copies of α are immediately adjacent to each other.

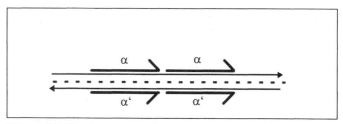

Figure 6.6 Tandem repeat primary structure

Tandem repeats of a substring $\omega \omega$ are not a context-free property.

Thus DNA and RNA cannot be adequately described with context-free languages.

Unbounded Reduplication

Unbounded reduplication means that a substring α may be repeated many times, and that there may even be gaps between repeated instances of the substring.

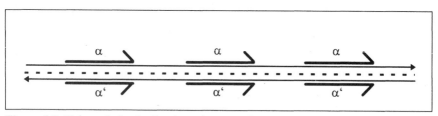

Figure 6.7 Unbounded reduplication primary structure

Indexed grammars provide a formalism to support unbounded reduplication [155, p. 193], but indexed grammars are a subclass of context-sensitive languages. Thus unbounded reduplication will require context-sensitive language capabilities. Indexed grammars will be studied in the next section.

tRNA Structures

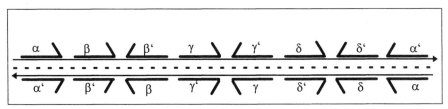

Figure 6.8 tRNA primary structure

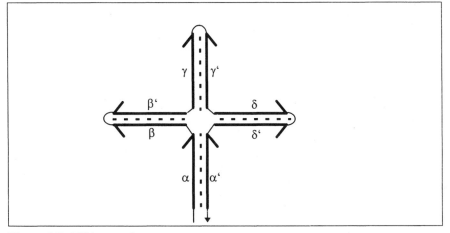

Figure 6.9 tRNA secondary structure

Thus we see that inverted repeats appear in tRNA secondary structures [155, p. 196], and we already have mentioned that inverted repeats are subsumed under context-free grammatical structures. Thus tRNA secondary structures also cannot be limited to regular expressions.

Multiple direct repeats are useful in dealing with mRNA attenuators [155, p. 197], which act as feedback factors. Once again, we find that context-free grammars are not adequate to express this language.

mRNA Attenuators

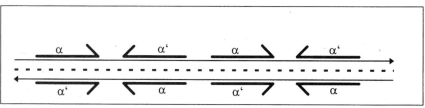

Figure 6.10 mRNA attenuator primary structure

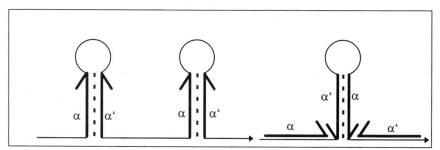

Figure 6.11 mRNA attenuator secondary structure

Superposition

Copia elements and tandem inverted repeats superimpose one kind of structure upon another structure, or are a combination of structures. Examples follow, but the point to be gained by studying these structures is that all may be supported by context-sensitive grammars, specifically indexed grammars.

Copia elements and tandem inverted repeats do not require power beyond context-sensitive grammars [155, pp. 198-200].

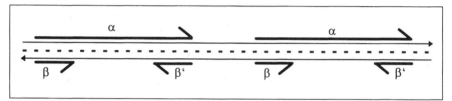

Figure 6.12 Copia elements, primary structure

The above structures can appear on either DNA or RNA strands.

Tandem Inverted Repeats

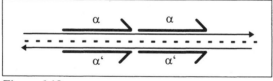

Figure 6.13

Tandem inverted repeats are just the same as the following.

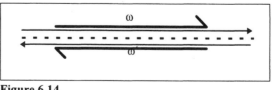

Figure 6.14

Mutation and Rearrangement [155, p. 202]

Point mutations are mutations that change a single base, while more extensive mutations may occur when a string of bases is replaced by another string of bases. Such replacement of terminal strings is supported by context-sensitive grammars, such as indexed grammars. Such replacements include deletions, insertions, and substitutions of strings of bases.

Excision [155, p. 203]

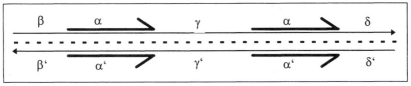

Figure 6.15

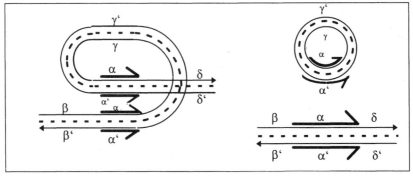

Figure 6.16

Inversion [155, p. 204]

A region in between inverted repeats becomes inverted.

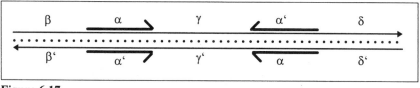

Figure 6.17

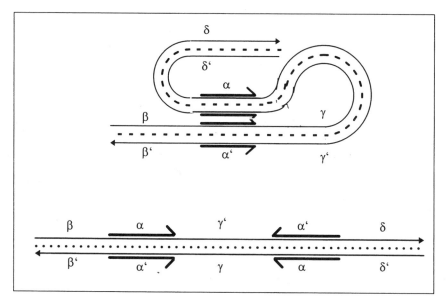

Figure 6.18

Recursive Secondary Structures

RNA may show instances of recursive structures within secondary structures [158, p. 74].

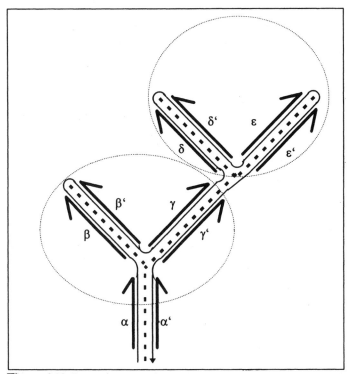

Figure 6.19 Recursive secondary structures

Pseudoknot Structure [158, p. 80]
(See Chapter 9 for further discussion about pseudoknots.)

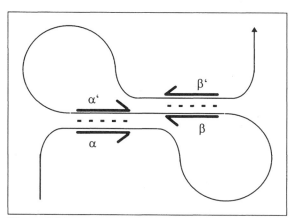

Figure 6.20

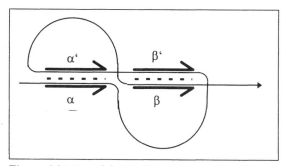

Figure 6.21 Pseudoknot with coaxial stacking [158, p. 95]

An example in English of a pseudoknot with respect to palindromes is the following (ignoring punctuation, spaces, and case differences) [160, p. 590].

"DNA's loops and spools" or, ignoring punctuation, spaces, and case differences:

dnasloopsandspools　　　where:　there are two palindromes:
　　　　　　　　　　　　　　　　　　dna　　　matches　　　and
　　　　　　　　　　　　　　　　　　sloops　matches　　　spools

Protein Languages [158, p. 102]

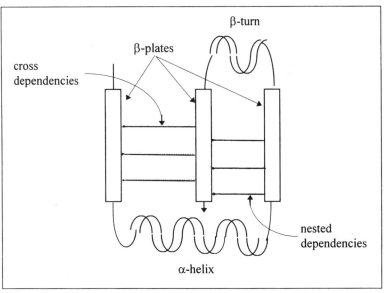

Figure 6.22 A hypothetical protein structure

If the above hypothetical protein structure is linearized into its primary structure, then the nested dependencies are characteristic of context-sensitive languages.

The conclusion to be drawn is that context-sensitive grammars, and specifically indexed grammars [155, p. 205] will adequately generate all the DNA and RNA structures discussed. As indexed grammars [161, p. 93] are a subset of context-sensitive grammars, they appear to be a tool that is adequate for the study of DNA, RNA, and proteins. It should be noted that Searls suggests that "...any grammar describing evolutionary change must in fact be greater than context-sensitive... The ability to create any number of duplications of arbitrary substrings on the input indicates that no linear-bounded automaton could recognize such a language, since these automata are limited to space which is linear in the size of the input... This does not mean that the strings of a language which is the product of an evolutionary grammar are necessarily greater than context-sensitive..." [158, pp. 110, 111]. Searls also notes that ambiguity is a factor to be considered, and applies to context-sensitive languages [166, example 2, p. 205] as much as to context-free languages.

Context-Sensitive Grammars and DNA

A proposed example of a context-sensitive grammar applied to DNA primary structure [158, p. 80]:

$$G^1 = \left(\{S, B_a, B_t, B_g, B_c\}, \{a, t, g, c\} \cup \{\bar{a}, \bar{t}, \bar{g}, \bar{c}\}, \mathcal{P}^1, S \right), \text{ where:}$$

$$\mathcal{P}^1 = \begin{cases}
S & \Rightarrow B_a aS \mid B_t tS \mid B_g gS \mid B_c cS \mid \lambda, \\
B_a & \Rightarrow \bar{a} \quad , \\
B_t & \Rightarrow \bar{t} \quad , \\
B_g & \Rightarrow \bar{g} \quad , \\
B_c & \Rightarrow \bar{c} \quad , \\
aB_a \Rightarrow B_a a \;, & aB_t \Rightarrow B_t a \;, & aB_g \Rightarrow B_g a \;, & aB_c \Rightarrow B_c a \\
tB_a \Rightarrow B_a t \;, & tB_t \Rightarrow B_t t \;, & tB_g \Rightarrow B_g t \;, & tB_c \Rightarrow B_c t \\
gB_a \Rightarrow B_a g \;, & gB_t \Rightarrow B_t g \;, & gB_g \Rightarrow B_g g \;, & gB_c \Rightarrow B_c g \\
cB_a \Rightarrow B_a c \;, & cB_t \Rightarrow B_t c \;, & cB_g \Rightarrow B_g c \;, & cB_c \Rightarrow B_c c
\end{cases}$$

When Watson-Crick complementary base relationships are invoked, we obtain the following context-sensitive grammar G^0:

<u>Watson-Crick Complementary Bases</u>

$$\bar{a} = t$$
$$\bar{t} = a$$
$$\bar{g} = c$$
$$\bar{c} = g$$

$$G^0 = \left(\{S, B_a, B_t, B_g, B_c\}, \{a, t, g, c\}, \mathcal{P}^0, S \right), \text{ where:}$$

$$\mathcal{P}^0 = \begin{cases}
S & \Rightarrow B_a aS \mid B_t tS \mid B_g gS \mid B_c cS \mid \lambda, \\
B_a & \Rightarrow t \quad , \\
B_t & \Rightarrow a \quad , \\
B_g & \Rightarrow c \quad , \\
B_c & \Rightarrow g \quad , \\
aB_a \Rightarrow B_a a \;, & aB_t \Rightarrow B_t a \;, & aB_g \Rightarrow B_g a \;, & aB_c \Rightarrow B_c a \\
tB_a \Rightarrow B_a t \;, & tB_t \Rightarrow B_t t \;, & tB_g \Rightarrow B_g t \;, & tB_c \Rightarrow B_c t \\
gB_a \Rightarrow B_a g \;, & gB_t \Rightarrow B_t g \;, & gB_g \Rightarrow B_g g \;, & gB_c \Rightarrow B_c g \\
cB_a \Rightarrow B_a c \;, & cB_t \Rightarrow B_t c \;, & cB_g \Rightarrow B_g c \;, & cB_c \Rightarrow B_c c
\end{cases}$$

As an example, we will generate a string, using grammar G^0.

$$S \Rightarrow B_a aS \Rightarrow B_a aB_g gS \Rightarrow B_a aB_g gB_t tS \Rightarrow B_a aB_g gB_t t \Rightarrow B_a B_g agB_t t$$
$$\Rightarrow B_a B_g aB_t gt \Rightarrow B_a B_g B_t agt \Rightarrow tB_g B_t agt \Rightarrow B_g tB_t agt \Rightarrow B_g B_t tagt$$
$$\Rightarrow cB_t tagt \Rightarrow B_t ctagt \Rightarrow act \cdot agt$$

Note.

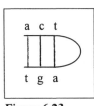

a c t

t g a

Figure 6.23

There is another set of languages, the languages generated by indexed grammars. Indexed grammar languages form a proper subset of context-sensitive languages, yet are not context-free. Indexed grammars are simpler to work with than unrestricted context-sensitive grammars, and will now be discussed.

Indexed Grammars [145, pp. 257, 258], [135, pp. 137-140]

An indexed grammar G is defined as follows.

$G = \left(V_N, V_T, \mathcal{P}, E, S \right)$ where: V_N is a set of non-terminal symbols

$\qquad V_T$ is a set of terminal symbols

$\qquad \mathcal{P}$ is a set of production rules of the form

$$X \Rightarrow (V_N \cdot E^* \cup V_T)^*$$

$\qquad E$ is a set of indices that identify

$\qquad$ index productions, $e_i: X \Rightarrow (V_N \cup V_T)^*$

Thus the right sides of production rules may contain non-terminals followed by arbitrarily long lists of indices. If a non-terminal followed by a list of indices is replaced by a derivation by one or more non-terminals, the list of indices then follows each of the new non-terminals.

Example: A $\Rightarrow$ a B e_1 e_3 e_4 d

and

B $\Rightarrow$ C j F H is used in the derivation,

then A $\Rightarrow$ a C e_1 e_3 e_4 j F e_1 e_3 e_4 H e_1 e_3 e_4 d

repeated list of indices
after each non-terminal
from the derivation

However, if a derivation is used with no non-terminals (terminals only), then the list of indices disappears.

Example 1: A $\Rightarrow$ a D e_2 e_4 B e_1 e_3 e_4 E e_4 e_5 d

and

B $\Rightarrow$ j is used in the derivation,

then A $\Rightarrow$ a D e_2 e_4 j E e_4 e_5 d

Note that the list of
indices e_1 e_3 e_4 is erased

Last rule: If a non-terminal X is followed by a list of indices e_1 e_2 e_3 ... e_n and e_1 names an index production rule X $\Rightarrow$ P (same X), then X is replaced by P, where each non-terminal in P is followed by the index list e_2 e_3 ... e_n and thus e_1 is consumed.

Q $\Rightarrow$ f A B e_1 e_2 e_3 d

and e_1 = { B $\Rightarrow$ D t K }

then

Q $\Rightarrow$ f A D e_2 e_3 t K e_2 e_3 d

Note: e_1 is erased

It is interesting that if $G = (V_N, V_T, P, E, S)$ is an indexed grammar and $E = \varnothing$, then G is a context-free grammar, but that in general, $L(G)$ when G is an indexed grammar is a context-sensitive grammar.

Examples of indexed grammars:

$$G_1 = (\{S, A, B\}, \{a, b, c\}, \{e_1, e_2\}, P_1, S),$$

where: $P_1 = \begin{cases} S & \Rightarrow & aAe_1c \\ A & \Rightarrow & aAe_2c \mid B \end{cases}$

and $\quad e_1 = \{ B \Rightarrow b \}$
$\quad\quad\quad e_2 = \{ B \Rightarrow bB \}$

Note: $L(G_1) = \{a^n b^n c^n \mid n > 0\}$ which is a well known context-sensitive language. A few derivations follow.

$S \Rightarrow aAe_1c \Rightarrow aBe_1c \Rightarrow abc$

$S \Rightarrow aAe_1c \Rightarrow aaAe_2e_1cc \Rightarrow aaBe_2e_1cc \Rightarrow aabBe_1cc \Rightarrow aabbcc$

$S \Rightarrow aAe_1c \Rightarrow aaAe_2e_1cc \Rightarrow aaaAe_2e_2e_1ccc \Rightarrow aaaBe_2e_2e_1ccc \Rightarrow aaabBe_2e_1ccc$
$\quad \Rightarrow aaabbBe_1ccc \Rightarrow aaabbbccc$

Another example:

$$G_2 = (\{S, T, A, B, C\}, \{a, b\}, \{f, g\}, P_2, S)$$

where: $P_2 = \begin{cases} S & \Rightarrow & Tf \\ T & \Rightarrow & Tg \mid ABA \end{cases}$

and $\quad f = \begin{cases} A & \Rightarrow & a \\ B & \Rightarrow & b \\ D & \Rightarrow & b \end{cases}, \quad g = \begin{cases} A & \Rightarrow & aA \\ B & \Rightarrow & bBCC \\ C & \Rightarrow & bC \end{cases}$

$$L(G_2) = \{a^n b^{n^2} a^n \mid n > 0\}$$

A few derivations follow:

$S \Rightarrow Tf \Rightarrow Af\ Bf\ Af \Rightarrow aBf\ Af \Rightarrow abAf \Rightarrow aba$

$S \Rightarrow Tf \Rightarrow Tgf \Rightarrow Agf\ Bgf\ Agf \Rightarrow aAf\ Bgf\ Agf \Rightarrow aAf\ bBf\ Cf\ Cf\ Agf$
$\Rightarrow aAf\ bBf\ Cf\ Cf\ aAf \Rightarrow aabBf\ Cf\ Cf\ aAf \Rightarrow aabbCf\ Cf\ aAf \Rightarrow aabbbCf\ aAf$
$\Rightarrow aabbbbaAf \Rightarrow a^2b^4a^2$

$S \Rightarrow Tf \Rightarrow Tgf \Rightarrow Tggf \Rightarrow Aggf\ Bggf\ Aggf \Rightarrow aAgf\ Bggf\ Aggf$
$\Rightarrow aaAf\ Bggf\ Agg \Rightarrow a^3\ Bggf\ Agg \Rightarrow a^3bBgf\ Cgf\ Cgf\ Aggf$
$\Rightarrow a^3b^2Bf\ Cf\ Cf\ Cgf\ Cgf\ Aggf \Rightarrow a^3b^3\ Cf\ Cf\ Cgf\ Cgf\ Aggf$
$\Rightarrow a^3b^3b\ Cf\ Cgf\ Cgf\ Aggf \Rightarrow a^3b^4b\ Cgf\ Cgf\ Aggf \Rightarrow a^3b^5b\ Cf\ Cgf\ Aggf$
$\Rightarrow a^3b^6b\ Cgf\ Aggf \Rightarrow a^3b^7b\ Cf\ Aggf \Rightarrow a^3b^8b\ Aggf \Rightarrow a^3b^9a\ Agf$
$\Rightarrow a^3b^9a^2\ Af \Rightarrow a^3b^9a^2a\ =\ a^3b^9a^3$

A slightly different notation may be used, in which the indices look like exponents [158, pp. 56, 57].

$$G = \left(\{S,\ T,\ A,\ B,\ C\},\ \{a,\ b,\ d\},\ \{s,\ t\},\ \mathcal{P},\ S \right)$$

  where: $\mathcal{P} = \{T \Rightarrow ABC\}$

  and

$$s = \begin{cases} S & \Rightarrow & T^s \\ A^s & \Rightarrow & a \\ B^s & \Rightarrow & b \\ C^s & \Rightarrow & d \end{cases}, \qquad t = \begin{cases} A^t & \Rightarrow & aA \\ B^t & \Rightarrow & bB \\ C^t & \Rightarrow & dC \\ T & \Rightarrow & T^t \end{cases}$$

A sample derivation using this notational variant of an indexed grammar:

$S \Rightarrow T^s \Rightarrow T^{ts} \Rightarrow T^{tts} \Rightarrow A^{tts}B^{tts}C^{tts} \Rightarrow aA^{ts}B^{tts}C^{tts} \Rightarrow aaA^sB^{tts}C^{tts}$
$\Rightarrow aaaB^{tts}C^{tts} \Rightarrow aaabB^{ts}C^{tts} \Rightarrow aaabbB^sC^{tts} \Rightarrow aaabbbC^{tts}$
$\Rightarrow aaabbbdC^{ts} \Rightarrow aaabbbddC^s \Rightarrow aaabbbddd$

Searls et al uses indexed grammars with this notation that looks like exponents to prove relationships between different systems of grammars. One of the problems is that reduplication (substrings of the form $\omega\omega$) implies context sensitivity, and thus the question being addressed is how can context-free languages be extended (into the class of context-sensitive languages) without requiring general context-sensitivity. Different grammars are studied, and the question being answered is what are the relationships between their corresponding languages.

We find [162, pp. 81-88] the following.

Adding additional index grammar rules such as the following proves containment relationships between different languages.

$$A \implies B^{ij} \quad , \quad A^{ij} \implies p \quad , \quad A \implies uB^i v$$

$$A^i \implies B^{ij} \quad , \quad A^{ij} \implies B^i \quad , \quad A \implies B^i C^{ij}$$

For example, to prove $A \implies B^{ij}$, we note the following:

$$A \implies C^j \quad \text{and} \quad C \implies B^i$$

and to prove $A \implies B^i C^{ij}$, we note the following:

$$A \implies DE \quad , \quad D \implies B^i \quad , \quad E \implies F^j \quad , \text{ and } \quad F \implies C^i$$

Relationship Between Grammars

Note the language containment relationships that result.

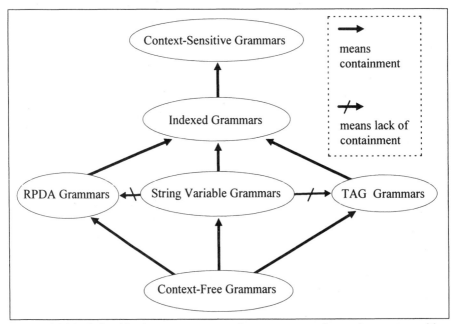

Figure 6.24 Relationships between grammars between context-free and context-sensitive

RPDA languages (Savitch, [148]) are generated by pushdown automata (thus are context-free), but have the additional context-sensitive capability of supporting nested duplicated strings of the form $\omega\alpha$

TAG (Tree-Adjoining Grammars) [191] are another attempt to extend context-free languages slightly into the context-sensitive domain, in an attempt to deal with natural-language capabilities by using non-transformational methods, thereby avoiding Chomsky's methods of dealing with natural languages.

Many people would consider Chomsky's transformational method of dealing with natural language to be somewhat adhoc. RPDA and TAG languages might be viewed as adhoc as well.

Definte Clause and String Variable Grammars

Most of the investigations concerning language by Searls et al have focused upon Definite Clause Grammars (DCG), and String Variable Grammars (SVG). From a practical standpoint, these methods focus upon the use of the programming language Prolog (an attempt to combine grammatical approaches with first-order predicate calculi and automatic theorem provers). The idea is that Prolog would also provide a method to extend the capabilities of context-free grammars.

It would appear that the best and cleanest characterization of all these investigations would result in index grammar languages (as a subset of context-free grammars) as being the best characterization of their investigations, and one upon which most investigators would agree.

Rather than dwell upon a philosophical discussion concerning natural languages, and whether RPDA, DCG, SVG, etc. are to be preferred, we shall leave this (perhaps sterile) discussion to the future.

An examination of some of the issues related to DNA, RNA, and proteins that have been successfully resolved by using index grammars will follow. First, let us apply indexed grammars to DNA [158, p. 81].

$$G^2 = \left(\{S, A\}, \{a, t, g, c\}, \{a, t, g, c\}, \mathcal{P}^2, S\right)$$

where:

$$\mathcal{P}^2 = \left\{\begin{matrix} S & \Rightarrow & A \\ A & \Rightarrow & \lambda \end{matrix}\right\}$$

and

$$a = \left\{\begin{matrix} S & \Rightarrow & aS^a \\ A^a & \Rightarrow & Aa \end{matrix}\right\}, \quad t = \left\{\begin{matrix} S & \Rightarrow & tS^t \\ A^t & \Rightarrow & At \end{matrix}\right\},$$

$$g = \left\{\begin{matrix} S & \Rightarrow & gS^g \\ A^g & \Rightarrow & Ag \end{matrix}\right\}, \quad c = \left\{\begin{matrix} S & \Rightarrow & cS^c \\ A^c & \Rightarrow & Ac \end{matrix}\right\}$$

As an example, let us generate a repeated string of the form $\omega\omega$

$$S \Rightarrow aS^a \Rightarrow agS^{ga} \Rightarrow agtS^{tga} \Rightarrow agtcS^{ctga} \Rightarrow agtcA^{ctga} \Rightarrow agtcA^{tga}c$$
$$\Rightarrow agtcA^{ga}tc \Rightarrow agtcA^{a}gtc \Rightarrow agtc \cdot agtc$$

Another indexed grammar [158, p. 81] can support substrings of the form $\omega \cdot \omega^R$.

$$G^3 = \left(\{S, A\}, \{a, t, g, c\} \cup \{\bar{a}, \bar{t}, \bar{g}, \bar{c}\}, \{a, t, g, c\}, \mathcal{P}^3, S \right)$$

where:

$$\mathcal{P}^3 = \begin{cases} S & \Rightarrow aS^a \mid tS^t \mid gS^g \mid cS^c \mid A \\ A & \Rightarrow \lambda \end{cases}$$

and

$$a = \{A^a \Rightarrow \bar{a}A\}, \qquad t = \{A^t \Rightarrow \bar{t}A\},$$

$$g = \{A^g \Rightarrow \bar{g}A\}, \qquad c = \{A^c \Rightarrow \bar{c}A\}$$

Using the Watson-Crick complementary bases,

$$\bar{a} = t, \quad \bar{t} = a$$
$$\bar{g} = c, \quad \bar{c} = g$$

we get grammar G^4.

$$G^4 = \left(\{S, A\}, \{a, t, g, c\}, \{a, t, g, c\}, \mathcal{P}^4, S \right)$$

where:

$$\mathcal{P}^4 = \begin{cases} S & \Rightarrow aS^a \mid tS^t \mid gS^g \mid cS^c \mid A \\ A & \Rightarrow \lambda \end{cases}$$

and

$$a = \{A^a \Rightarrow tA\}, \qquad t = \{A^t \Rightarrow aA\},$$

$$g = \{A^g \Rightarrow cA\}, \qquad c = \{A^c \Rightarrow gA\}$$

An example of a derivation:

$$S \Rightarrow aS^a \Rightarrow atS^{ta} \Rightarrow attS^{tta} \Rightarrow attgS^{gtta} \Rightarrow attgcS^{cgtta} \Rightarrow attgcA^{cgtta}$$
$$\Rightarrow attgcgA^{gtta} \Rightarrow attgcgcA^{tta} \Rightarrow attgcgcaA^{ta} \Rightarrow attgcgcaaA^{a}$$
$$\Rightarrow attgc \cdot gcaat$$

The next example will be an indexed grammar that allows "interleaved" repeats [158, p. 82].

$$G^5 = \left(\{S, A\}, \{a, t, g, c\} \cup \{\bar{a}, \bar{t}, \bar{g}, \bar{c}\}, \{a, t, g, c\}, \mathcal{P}^5, S \right)$$

where:

$$\mathcal{P}^5 = \begin{cases} S & \Rightarrow & aS^a \mid tS^t \mid gS^g \mid cS^c \mid A \\ B & \Rightarrow & \bar{a}B \mid \bar{t}B \mid \bar{g}B \mid \bar{c}B \mid \lambda \end{cases}$$

and

$$a = \left\{ A^a \Rightarrow aA\bar{a} \mid B \right\}, \qquad t = \left\{ A^t \Rightarrow tA\bar{t} \mid B \right\},$$

$$g = \left\{ A^g \Rightarrow gA\bar{g} \mid B \right\}, \qquad c = \left\{ A^c \Rightarrow cA\bar{c} \mid B \right\}$$

Using the Watson-Crick complementary bases,

$$\begin{aligned} \bar{a} &= t, & \bar{t} &= a \\ \bar{g} &= c, & \bar{c} &= g \end{aligned}$$

we get the following grammar.

$$G^6 = \left(\{S, A\}, \{a, t, g, c\}, \{a, t, g, c\}, \mathcal{P}^6, S \right)$$

where:

$$\mathcal{P}^6 = \begin{cases} S & \Rightarrow & aS^a \mid tS^t \mid gS^g \mid cS^c \mid A \\ B & \Rightarrow & tB \mid aB \mid cB \mid gB \mid 1 \end{cases}$$

and

$$a = \left\{ A^a \Rightarrow aAt \mid B \right\}, \qquad t = \left\{ A^t \Rightarrow tAa \mid B \right\},$$

$$g = \left\{ A^g \Rightarrow gAc \mid B \right\}, \qquad c = \left\{ A^c \Rightarrow cAg \mid B \right\}$$

A sample derivation follows.

$S \Rightarrow aS^a \Rightarrow atS^{ta} \Rightarrow attS^{tta} \Rightarrow attgS^{gtta} \Rightarrow attgcS^{cgtta} \Rightarrow attgcA^{cgtta}$

$\Rightarrow attgc \cdot cA^{gtta}g \Rightarrow attgc \cdot cgA^{tta}cg \Rightarrow attgc \cdot cgtA^{ta}acg \Rightarrow attgc \cdot cgttA^a aacg$

$\Rightarrow attgc \cdot cgttAtaacg \Rightarrow attgc \cdot cgttaBtaacg \Rightarrow attgc \cdot cgtta \cdot taacg$

Other sublanguages or subgrammars have been developed that are explicitly context-sensitive. By "sublanguage" or "subgrammar" is meant a language or grammar intended to deal with a specific problem, ignoring other aspects of the language or grammar, in this case, dealing with all problems of DNA (or RNA or Proteins).

Direct Repeat Context-Sensitive Grammar

A context-sensitive grammar to support direct repeats [160, p. 589]:

$$G^{dr} = \left(\{S, \ A, \ C, \ G, \ T, \ X\}, \ \{a, \ c, \ g, \ t\}, \ \boldsymbol{P}^{dr}, \ S \right),$$

where: $\boldsymbol{P}^{dr}$ is as follows.

S	$\Rightarrow$	AaS | CcS | GgS | TtS | X							
X	$\Rightarrow$	λ							
Aa	$\Rightarrow$ aA	,	Ca	$\Rightarrow$ aC	,	Ga	$\Rightarrow$ aG	,	Ta $\Rightarrow$ aT
Ac	$\Rightarrow$ cA	,	Cc	$\Rightarrow$ cC	,	Gc	$\Rightarrow$ cG	,	Tc $\Rightarrow$ cT
Ag	$\Rightarrow$ gA	,	Cg	$\Rightarrow$ gC	,	Gg	$\Rightarrow$ gG	,	Tg $\Rightarrow$ gT
At	$\Rightarrow$ tA	,	Ct	$\Rightarrow$ tC	,	Gt	$\Rightarrow$ tG	,	Tt $\Rightarrow$ tT
AX	$\Rightarrow$ Xa	,	CX	$\Rightarrow$ Xc	,	GX	$\Rightarrow$ Xg	,	TX $\Rightarrow$ Xt

A sample derivation.

$$
\begin{aligned}
S \Rightarrow{} & CcS \Rightarrow CcTtS \Rightarrow CcTtAaS \Rightarrow CcTtAaAaS \Rightarrow CcTtAaAaCcS \\
\Rightarrow{} & cCTtAaAaCcS \Rightarrow cCtTAaAaCcS \Rightarrow ctCTAaAaCcS \\
\Rightarrow{} & ctCTaAAaCcS \Rightarrow ctCaTAAaCcS \Rightarrow ctaCTAAaCcS \\
\Rightarrow{} & ctaCTAaACcS \Rightarrow ctaCTaAACcS \Rightarrow ctaCaTAACcS \\
\Rightarrow{} & ctaaCTAACcS \Rightarrow ctaaCTAAcCS \Rightarrow ctaaCTAcACS \\
\Rightarrow{} & ctaaCTcAACS \Rightarrow ctaaCcTAACS \Rightarrow ctaacCTAACS \\
\Rightarrow{} & ctaacCTAACX \Rightarrow ctaacCTAAXc \Rightarrow ctaacCTAXac \\
\Rightarrow{} & ctaacCTXaac \Rightarrow ctaacCXtaac \Rightarrow ctaacXctaac \Rightarrow \\
\Rightarrow{} & ctaac\,\lambda\,ctaac = ctaac\cdot ctaac
\end{aligned}
$$

Pseudoknot with Direct Repeat Grammar [160, p. 589]

$$G^{Pkdr} = \left(\{S, \ A, \ C, \ G, \ T, \ P, \ Q, \ X\}, \ \{a, \ c, \ g, \ t\}, \ \mathcal{P}^{Pkdr}, \ S \right)$$

where: $\mathcal{P}^{Pkdr}$ is as follows.

$$
\begin{aligned}
S &\Rightarrow AaS \mid CcS \mid GgS \mid TtS \mid PX \mid X \\
X &\Rightarrow \lambda \\
P &\Rightarrow aPt \mid cPg \mid gPc \mid tPa \mid Q \\
Q &\Rightarrow AQt \mid CQg \mid GQc \mid TQa \mid \lambda
\end{aligned}
$$

Aa $\Rightarrow$ aA ,	Ca $\Rightarrow$ aC ,	Ga $\Rightarrow$ aG ,	Ta $\Rightarrow$ aT
Ac $\Rightarrow$ cA ,	Cc $\Rightarrow$ cC ,	Gc $\Rightarrow$ cG ,	Tc $\Rightarrow$ cT
Ag $\Rightarrow$ gA ,	Cg $\Rightarrow$ gC ,	Gg $\Rightarrow$ gG ,	Tg $\Rightarrow$ gT
At $\Rightarrow$ tA ,	Ct $\Rightarrow$ tC ,	Gt $\Rightarrow$ tG ,	Tt $\Rightarrow$ tT
AX $\Rightarrow$ Xa ,	CX $\Rightarrow$ Xc ,	GX $\Rightarrow$ Xg ,	TX $\Rightarrow$ Xt

A sample derivation.

$$
\begin{aligned}
S \Rightarrow{} & PX \Rightarrow gPcX \Rightarrow gcPgcX \Rightarrow gcaPtgcX \Rightarrow gcagPctgcX \Rightarrow gcagQctgcX \\
\Rightarrow{} & gcagCQgctgcX \Rightarrow gcagCAQtgctgcX \Rightarrow gcagCATQatgctgcX \Rightarrow \\
\Rightarrow{} & gcagCATTQaatgctgcX \Rightarrow gcagCATT\,\lambda\,aatgctgcX = gcagCATTaatgctgcX \\
\Rightarrow{} & gcagCATaTatgctgcX \Rightarrow gcagCAaTTatgctgcX \Rightarrow gcagCaATTatgctgcX \\
\Rightarrow{} & gcagaCATTatgctgcX \Rightarrow gcagaCATaTtgctgcX \Rightarrow gcagaCAaTTtgctgcX \\
\Rightarrow{} & gcagaCaATTtgctgcX \Rightarrow gcagaaCATTtgctgcX \Rightarrow gcagaaCATtTgctgcX \\
\Rightarrow{} & gcagaaCAtTTgctgcX \Rightarrow gcagaaCtATTgctgcX \Rightarrow gcagaatCATTgctgcX \\
\Rightarrow{} & gcagaatCATgTctgcX \Rightarrow gcagaatCAgTTctgcX \Rightarrow gcagaatCgATTctgcX \\
\Rightarrow{} & gcagaatgCATTctgcX \Rightarrow gcagaatgCATcTtgcX \Rightarrow gcagaatgCAcTTtgcX \\
\Rightarrow{} & gcagaatgCcATTtgcX \Rightarrow gcagaatgcCATTtgcX \Rightarrow gcagaatgcCATtTgcX \\
\Rightarrow{} & gcagaatgcCATtTgcX \Rightarrow gcagaatgcCAtTTgcX \Rightarrow gcagaatgcCtATTgcX \\
\Rightarrow{} & gcagaatgctCATTgcX \Rightarrow gcagaatgctCATgTcX \Rightarrow gcagaatgctCAgTTcX \\
\Rightarrow{} & gcagaatgctCgATTcX \Rightarrow gcagaatgctgCATTcX \Rightarrow gcagaatgctgCATcTX \\
\Rightarrow{} & gcagaatgctgCAcTTX \Rightarrow gcagaatgctgCcATTX \Rightarrow gcagaatgctgcCATTX \\
\Rightarrow{} & gcagaatgctgcCATXt \Rightarrow gcagaatgctgcCAXtt \Rightarrow gcagaatgctgcCXatt \\
\Rightarrow{} & gcagaatgctgcXcatt \Rightarrow gcagaatgctgclcatt = gcagaatgctgccatt
\end{aligned}
$$

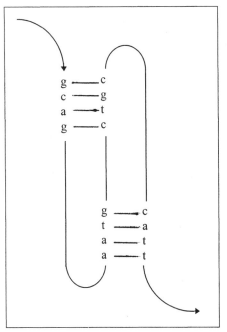

Figure 6.25   Secondary structure of RNA showing the pseudoknot with direct repeats

Chomsky-Schützenberger Theorem Revisited

When discussing context-free languages, we noted that many linguistic elements of DNA and RNA structures could be supported by languages which in fact were Dyck languages. There is yet more to this viewpoint. There is a celebrated theorem that extends the Chomsky-Schützenberger Theorem.

$$L_0 = h\left(L_2^1 \cap L_2^2\right),$$

where: $\left\{L_2^i\right\}_{i=1,2}$ are context-free (type 2 languages),

h is a homomorphism

L_0 is type 0 language

Thus context-free languages to support various factors such as palindromic structures can be intersected, and under homomorphism, are type 0. Of course, context-sensitive languages are a subset of type 0 languages.

Replication Fork Context-Sensitive Grammars

Recall that cut grammars [164] were discussed in a previous chapters were limited to Chomsky type 3 regular grammars and type 2 context-free grammars, with their finite state automata and pushdown automata, respectively. At that time, discussions of grammars relating to context-sensitive grammars was postponed. We continue the discussion of context-free grammars with specific applications. This discussion will include grammars similar to cut grammars such as indexed grammars, and we continue the discussion.

Let us examine the replication fork (which also helps explain the Holliday structure, Figure 5.13). This discussion is taken from Searls [164, p.p. 34-35]. As the DNA strands can only be copied in one direction (the direction of the arrows), the lower strand must be copied in short fragments referred to as Okazaki fragments. The extra nick allows short fragments, and may later be ligated. Thus the S and A strands must be identical (strands derived from B are independent). Such a duplicated copy implies a non context-free type language (at least context-sensitive). See Figure 6.29.

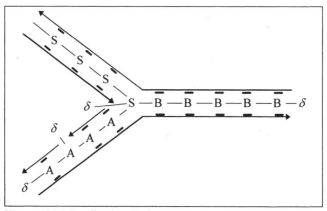

Figure 6.26 Replication Fork

An indexed grammar for the replication fork structure follows.

$$G_{rep\ fork} = \left(\{S,\ A,\ B\},\ \{a,\ c,\ g,\ t\},\ \mathcal{P}_{rep\ fork},\ S \right),\ where:$$

$$\mathcal{P}_{rep\ fork} = \begin{cases} S & \Rightarrow & aS^a t \mid cS^c g \mid gS^g c \mid tS^t a \mid \delta\,AB \\ A^a & \Rightarrow & tAa \\ A^c & \Rightarrow & gAc \\ A^g & \Rightarrow & cAg \\ A^t & \Rightarrow & aAt \\ A & \Rightarrow & \delta\,A \mid \delta \\ B & \Rightarrow & aBt \mid cBg \mid gBc \mid tBa \mid \delta \end{cases}$$

A sample derivation follows.[a]

$$
\begin{aligned}
S &\Rightarrow tS^t a \\
&\Rightarrow taS^{at} ta \\
&\Rightarrow tacS^{cat} gta \\
&\Rightarrow tacgS^{gcat} cgta \\
&\Rightarrow tacg\,\delta\,A^{gcat}\,Bcgta \\
&\Rightarrow tacg\,\delta\,cA^{cat}\,g\,Bcgta \\
&\Rightarrow tacg\,\delta\,c\,gA^{at}\,cg\,Bcgta \\
&\Rightarrow tacg\,\delta\,cg\,tA^{t}\,acgBcgta \\
&\Rightarrow tacg\,\delta\,cgt\,\delta\,A^{t}acgBcgta \\
&\Rightarrow tacg\,\delta\,cgt\,\delta\,aA\ \ tacgBcgta \\
&\Rightarrow tacg\,\delta\,cgt\,\delta\,a\,\delta\ \ tacg\cdot B\cdot cgta \\
&\Rightarrow tacg\,\delta\,cgt\,\delta\,a\,\delta\ \ tacg\cdot tBa\cdot cgta \\
&\Rightarrow tacg\,\delta\,cgt\,\delta\,a\,\delta\ \ tacg\cdot t\,aBta\cdot cgta \\
&\Rightarrow tacg\,\delta\,cgt\,\delta\,a\,\delta\ \ tacg\cdot ta\,gBcta\cdot cgta \\
&\Rightarrow tacg\,\delta\,cgt\,\delta\,a\,\delta\ \ tacg\cdot tag\,tBacta\cdot cgta \\
&\Rightarrow tacg\,\delta\,cgt\,\delta\,a\,\delta\ \ tacg\cdot tagt\,\delta\,acta\cdot cgta
\end{aligned}
$$

----

[a] $S \Rightarrow CB$ and $C \Rightarrow A^i$ thus $S \Rightarrow A^i B$, thus if $T^j \Rightarrow W^j V$ and $W \Rightarrow U^i$, then $T^j \Rightarrow U^{ij} V$. In this case, $S^{gcat} \Rightarrow \delta\,A^{gcat}\,B$.

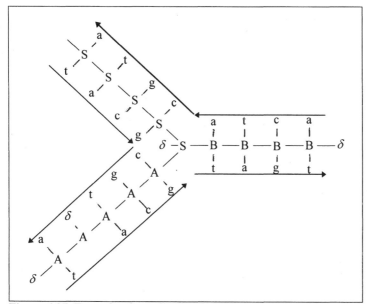

Figure 6.27 Example of a replication fork using a context-sensitive indexed grammar

Ligated languages [164, pp. 35 - 37] will be the last cut-grammar like method to be discussed and will only be discussed briefly.

A ligated grammar is like a cut grammar (review Chapter 5), but has an additional symbol $\gamma \notin V_N \cup V_T$ and $\gamma \neq \delta$.

If $\omega = \omega_1 \gamma \omega_2 \gamma \omega_3 \dots \omega_{n-1} \gamma \omega_n \gamma$, where: $\omega_i \in (V_T \cup \delta)^*$,

   and $\tilde{\omega} = \{\tilde{\omega}_1, \tilde{\omega}_2, \dots, \tilde{\omega}_n\}$

then $\bar{L}(G) = \{\tilde{\omega} \mid \tilde{\omega} \in (2^{V_T})^* \text{ and } S \overset{*}{\Rightarrow} \omega\}$.

Thus the ligation language combines different numbers of $\tilde{\omega}_i$ in different orders, so that the resulting strings are effectively ligated. The ligations only take place at the δ gaps at the gaps at the γ symbols remain (ligations cannot happen anywhere, only at δ sites).

An example of a ligation grammar follows.

$G = (\{S, A, B\}, \{a, c, g, t\}, \mathcal{P}, S)$

where: Watson-Crick complements

$$\bar{a} = t$$
$$\bar{c} = g$$
$$\bar{g} = c$$
$$\bar{t} = a$$

and $\mathcal{P} = \begin{cases} S & \Rightarrow & aSt \mid cSg \mid gSc \mid tSa \mid \delta A \mid \delta B \mid \gamma \\ A & \Rightarrow & \gamma\, Aa \mid \gamma\, Ac \mid \gamma\, Ag \mid \gamma\, At \mid aSt \mid cSg \mid gSc \mid tSa \mid \gamma \\ B & \Rightarrow & aB\gamma \mid cB\gamma \mid gB\lambda \mid tB\gamma \mid aSt \mid cSg \mid gSc \mid tSa \mid \gamma \end{cases}$

While this grammar is context-free, it never the less extends our repertoire to ligation.

As a sample derivation, we have the following.

$$
\begin{aligned}
S & \Rightarrow aSt \\
& \Rightarrow acSgt \\
& \Rightarrow actSagt \\
& \Rightarrow act\,\delta\, Aagt \\
& \Rightarrow act\,\delta\gamma\, Agagt \\
& \Rightarrow act\,\delta\gamma\, aStgagt \\
& \Rightarrow act\,\delta\gamma\, agSctgagt \\
& \Rightarrow act\,\delta\gamma\, agaStctgagt \\
& \Rightarrow act\,\delta\gamma\, agaB\,\delta\, tctgagt \\
& \Rightarrow act\,\delta\gamma\, agatSa\,\delta\, tctgagt \\
& \Rightarrow act\,\delta\gamma\, \underbrace{agat}_{\tilde{\omega}_1}\, \gamma\, \underbrace{a}_{\tilde{\omega}_2}\, \delta\, \underbrace{tctgagt}_{\tilde{\omega}_3}
\end{aligned}
$$

Thus $\bar{L} = \begin{cases} \lambda, \\ \tilde{\omega}_1,\ \tilde{\omega}_2,\ \tilde{\omega}_3, \\ \tilde{\omega}_1\tilde{\omega}_2,\ \tilde{\omega}_2\tilde{\omega}_1,\ \tilde{\omega}_1\tilde{\omega}_3,\ \tilde{\omega}_3\tilde{\omega}_1,\ \tilde{\omega}_2\tilde{\omega}_3,\ \tilde{\omega}_3\tilde{\omega}_2, \\ \tilde{\omega}_1\tilde{\omega}_2\tilde{\omega}_3,\ \tilde{\omega}_1\tilde{\omega}_3\tilde{\omega}_2,\ \tilde{\omega}_2\tilde{\omega}_1\tilde{\omega}_3,\ \tilde{\omega}_2\tilde{\omega}_3\tilde{\omega}_1,\ \tilde{\omega}_3\tilde{\omega}_1\tilde{\omega}_2,\ \tilde{\omega}_3\tilde{\omega}_2\tilde{\omega}_1 \end{cases}$

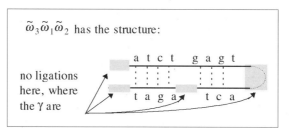

$\tilde{\omega}_3\tilde{\omega}_1\tilde{\omega}_2$ has the structure:

no ligations here, where the γ are

a t c t   g a g t

t a g a   t c a

Figure 6.28 Language of ligated strings

Context-Sensitive Grammars and Gene Regulation

What is significant in this discusion is the construction of a context-sensitive grammar in which the sets of non-terminals and terminals act as meta symbols. With the vocabulary set being meta variables, subgrammars (all context-sensitive) are constructed that have biological significance in the area of DNA, operons, etc.!

Abbreviating to ease comprehension, we find the following context-sensitive grammar [21, p. 340].

Non-Terminals			Terminals		
INDUCTION	~	A	*repressor_ cf _site*	~	a
EXPRESSION	~	B	*inducer*	~	b
DIRECT_ACTIVATION	~	C	*activator_ cf _site*	~	c
ACTIVATION	~	D	*inhibitor*	~	d
INHIBITION	~	E	*repressor_ dna_ site*	~	e
REPRESSION	~	F	*co − repressor*	~	f
DIRECT_REPRESSION	~	H	*activator_ dna_ site*	~	g
operator	~	I	*activator_ ac_ site*	~	h
DNAsiteRE	~	J	*RNApolymerase*	~	i
INITIATION_COMPLEX	~	K	*RNApolymerase_ site*	~	j
promotor	~	L	*newProtein*	~	k

$$G^0 = \left(V_N^0, \ V_T^0, \ \mathcal{P}^0, \ S\right) \text{where} \quad V_N^0 = \left\{S, \ A, \ B, \ C, \ D, \ E, \ F, \ H, \ I, \ J, \ K, \ L\right\}$$

$$\text{and} \quad V_T^0 = \left\{a, \ b, \ c, \ d, \ e, \ f, \ g, \ h, \ i, \ j, \ k\right\}$$

the production rules are as follows.

$\mathcal{P}^0$:

1	S	$\Rightarrow$	ACS | ES | FS | BS | CS | DS | ABS | ADS | HS | I
2	A	$\Rightarrow$	ab
3	E	$\Rightarrow$	cd
4	H	$\Rightarrow$	eI
5	F	$\Rightarrow$	afI
6	afI	$\Rightarrow$	efI
7	C	$\Rightarrow$	gJK
8	D	$\Rightarrow$	cbJK
9	cbJ	$\Rightarrow$	gbJ
10	gbJ	$\Rightarrow$	hbJ
11	gJ	$\Rightarrow$	hJ
12	K	$\Rightarrow$	iL
13	hbJiL	$\Rightarrow$	hbJjL
14	hJiL	$\Rightarrow$	hJjL
15	B	$\Rightarrow$	jL
16	jL	$\Rightarrow$	jLk

A second context-sensitive subgrammar is then constructed [21, p. 341].

Non-Terminals:			Terminals:		
I_1	~	*lac_operator*	a_1	~	*lac_repressor_cf_site*
			b_1	~	*lactose*
			b_2	~	*cAMP*
			c_1	~	*cap_cf_site*
			e_1	~	*lac_repressor_dna_site*
			g_1	~	*cap_dna_site*
			h_1	~	*cap_ac_site*
			k_1	~	*lac_enzymes*

$$G^1 = \left(V_N^1,\; V_T^1,\; \mathcal{P}^1,\; S\right) \quad \text{where } V_N^1 = \left\{S,\; A,\; B,\; D,\; H,\; I_1,\; J,\; K,\; L\right\},$$

$$\text{and} \quad V_T^1 = \left\{a_1,\; b_1,\; b_2,\; c_1,\; e_1,\; g_1,\; h_1,\; i,\; j,\; k_1\right\}$$

the production rules are as follows.

$\mathcal{P}^1$:

1	S	$\Rightarrow$	ABS | ADS | HS | 1
2	A	$\Rightarrow$	$a_1 b_1$
4	H	$\Rightarrow$	$e_1 I_1$
8	D	$\Rightarrow$	$c_1 b_2 JK$
9	$c_1 b_2 J$	$\Rightarrow$	$g_1 b_2 J$
10	$g_1 b_2 J$	$\Rightarrow$	$h_1 b_2 J$
12	K	$\Rightarrow$	iL
13	$h_1 b_2 JiL$	$\Rightarrow$	$h_1 b_2 JjL$
15	B	$\Rightarrow$	jL
16	jL	$\Rightarrow$	jLk_1

A third context-sensitive subgrammar is then constructed [21, p. 343].

Non-Terminals			Terminals		
J_1	~	*MREsite*	c_2	~	*receptor_ cf _site*
J_2	~	*GREsite*	b_3	~	*hormone*
			g_2	~	*heavyMetal*
			g_3	~	*receptor_ dna _site*
			h_2	~	*heavyMetal_ ac _site*
			h_3	~	*receptor_ ac _site*
			k_2	~	*metallothionein*

$$G^2 = \left(V_N^2, \ V_T^2, \ \mathcal{P}^2, \ S\right)$$

where $V_N^2 = \left\{S, \ B, \ C, \ D, \ J_1, \ J_2, \ K, \ L\right\}$

and $V_T^2 = \left\{b_3, \ c_2, \ g_2, \ g_3, \ h_2, \ h_3, \ i, \ j, \ k_2\right\}$

the production rules are as follows.

$\mathcal{P}^2$:

1	S	$\Rightarrow$	BS	CS	DS	1
7	C	$\Rightarrow$	$g_2 J_1 K$
8	D	$\Rightarrow$	$c_2 b_3 J_2 K$
9	$c_2 b_3 J_2$	$\Rightarrow$	$g_3 b_3 J_2$
10	$g_3 b_3 J_2$	$\Rightarrow$	$h_3 b_3 J_2$
11	$g_2 J_1$	$\Rightarrow$	$h_2 J_1$
12	K	$\Rightarrow$	iL
13	$h_3 b_3 J_2 iL$	$\Rightarrow$	$h_3 b_3 J_2 jL$
14	$h_2 J_1 iL$	$\Rightarrow$	$h_2 J_1 jL$
15	B	$\Rightarrow$	jL
16	jL	$\Rightarrow$	jLk_2

A fourth context-sensitive subgrammar is then constructed [21, p. 343]:

Non-Terminals:		Terminals:	
I_2 $\sim$	DNAsite_851_881	a_2 $\sim$	GAL80_cf_site
J_3 $\sim$	DNAsite_1_65	b_4 $\sim$	galactose
		d_1 $\sim$	GAL80
		e_2 $\sim$	GAL80_dna_site
		e_3 $\sim$	GAL4_cf_site
		g_4 $\sim$	GAL4_dna_site
		h_4 $\sim$	GAL4_ac_site

$$G^3 = \left(V_N^3, \ V_T^3, \ \mathcal{P}^3, \ S \right)$$

where: $V_N^3 = \left\{ S, \ A, \ C, \ E, \ H, \ I_2, \ J_3, \ K, \ L \right\}$

and $V_T^3 = \left\{ a_2, \ b_4, \ d_1, \ e_2, \ e_3, \ g_4, \ h_4, \ i, \ j, \ k \right\}$

the production rules are as follows.

$\mathcal{P}^3$:

1	S	$\Rightarrow$	ACS | ES | HS | l
2	A	$\Rightarrow$	$a_2 b_4$
3	E	$\Rightarrow$	$e_3 d_1$
4	H	$\Rightarrow$	$e_2 I_2$
7	C	$\Rightarrow$	$g_4 J_3 K$
11	$g_4 J_3$	$\Rightarrow$	$h_4 J_3$
12	K	$\Rightarrow$	iL
14	$h_4 J_3 iL$	$\Rightarrow$	$h_4 J_3 jL$
16	jL	$\Rightarrow$	jLk

A final fifth context-sensitive subgrammar is then constructed [21, p. 344]:

Terminals		
f_1	$\sim$	*tryptophan*
k_3	$\sim$	*trp_enzyme*

$$G^4 = \left(\left\{ S, \ B, \ F, \ I, \ L \right\}, \ \left\{ a, \ e, \ f_1, \ j, \ k_3 \right\}, \ \mathcal{P}^4, \ S \right)$$

where production rules are as follows.

$\mathcal{P}^4$:

1	S	$\Rightarrow$	BS | FS | l
5	F	$\Rightarrow$	$af_1 I$
6	$af_1 I$	$\Rightarrow$	$ef_1 I$
15	B	$\Rightarrow$	jL
16	jL	$\Rightarrow$	jLk_3

If we compare context-sensitive grammars $\left\{G^i\right\}_{i=0}^{4}$ we find that the context-sensitive

grammar G^0 generates all the other context-sensitive grammars $\left\{G^i\right\}_{i=1}^{4}$ if meta-non-

terminals and meta-terminals are used.

Meta-Non-Terminals	Meta-Terminals

$\mathscr{I} \sim I , I_1, I_2$

$\mathscr{J} \sim J , J_1, J_2, J_3$

$a \sim a , a_1, a_2$

$b \sim b , b_1, b_2, b_3, b_4$

$c \sim c , c_1, c_2$

$d \sim d , d_1$

$e \sim e , e_1, e_2, e_3$

$f \sim f , f_1$

$g \sim g , g_1, g_2, g_3, g_4$

$h \sim h , h_1, h_2, h_3, h_4$

$k \sim k , k_1, k_2, k_3$

Context-Sensitive Grammars for Cytochrome-C

A context-sensitive grammar has been constructed that describes the cytochrome-c protein for many life forms [70, pp. 56-58]. This grammar describes cytochrome-c for animals after the ancestor of all animals in the phylogenetic tree below.

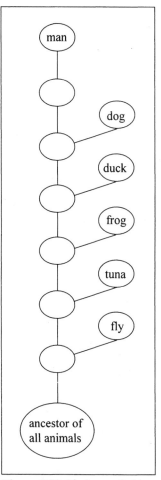

Figure 6.29 Phylogenetic hierarchy based upon cytochrome-c

$$G = \left(\{S, S_1, q_2, q_3\} \cup \{\chi_n\}_{n=1}^{8} \cup \{\chi_m\}_{m=9}^{17} \cup \{a, b, c, d, e, f\}, V_T, \mathcal{P}, S\right)$$

where $V_T = \{A, C, D, E, F, G, H, I, K, L, M, N, P, Q, R, S, T, V, W, Y\}$

and the set of production rules $\mathcal{P}$ is as follows.

$S \quad \Rightarrow \chi_n G \chi_4 G \chi_7 C \chi_2 CH \chi_8 K \chi_1 G_p \chi_1 L \chi_1 G \chi_{13} Y \chi_{10} W \chi_7 Y \chi_2 NPK \chi_6 KM \chi_1 F \chi_4 K \chi_m$

$K \chi_1 G \Rightarrow KA_{1,i} G$

$A_{1,i} \Rightarrow T \mid V$

$P \chi_1 L \Rightarrow PNL$

$L \chi_1 G \Rightarrow LA_{2,i} G$

$A_{2,i} \Rightarrow H \mid N \mid Y \mid W$

$M \chi_1 F \Rightarrow MA_{3,i} F$

$A_{3,i} \Rightarrow I \mid V$

$Y \chi_2 N \Rightarrow YLEN$

$C \chi_2 C \Rightarrow C q_2 C$

$q_2 \quad \Rightarrow SQ \mid AQ$

$F \chi_4 K \Rightarrow FrKK$

$r \quad \Rightarrow VGI \mid AGI \mid AGL$

(continued)

$G \chi_4 G$ ⇑ Gq_3G

q_3 ⇑ DVEK | DIEK | DVAK | NAEN

$K \chi_6 K$ ⇑ KKYIPGTK

$G \chi_7 C$ ⇑ $GKKS_1C$

S_1 ⇑ IFIMK | IFVQK | TFVQK | IFVQR

$W \chi_7 Y$ ⇑ WtTLuY

t ⇑ GED | KEE | GEE | GED | NND | QDD | GDD

u ⇑ ME | FE

$H \chi_8 K$ ⇑ HTvGGKH

v ⇑ VEK | CEK | VEN | VEA

$Y \chi_{10} W$ ⇑ YwNKxKGI χ_1 W

χ_3 ⇑ TAA | TDA | TEA | TNA | SNA

x ⇑ N | S | A

y ⇑ I | T | V

$G \chi_{13} Y$ ⇑ GaGRKTGQAbGcY

a ⇑ LF | IF | LI | FY

b ⇑ P | V | E | A

c ⇑ YS | FT | FS | FA

$e \chi_n$ ⇑ e | eGVPA

$\chi_m e$ ⇑ cERdLeAYLKfe

c ⇑ KE | KT | KG | TG | KD | KS | KA | PN | AN

d ⇑ AD | ED | VD | QD | GD

e ⇑ I | V

f ⇑ KATNE | KATKE | DATSK | DATAK | SACSK | SATS | SATK | ESTK

Chapter 7: Turing Machines and Sub-Turing Machines

Names of Central and South American Epidemic Diseases [a]	
calenturas cuartanas	malaria (chills every four days)
calenturas tercianas	malaria (chills every three days)
catarro	influenza
cocoliztli	plague
disentería de sangre	bloody dysentery
dolor de costado	chest or side pains
escarlatina, scarlatina	scarlet fever
garrotillo	diptheria
gucumatz, k'ucumatz	pneumonic plague
hueycocoliztli	pestilence
mal de pujos	dysentery
matlaltotonqui, matlalzahuatl, tabardillo	typhus
paperas, quechpozahualiztli	mumps
peste	epidemic
pian	verruga peruana (Carrion's disease)
pinto	spirochete disease
romadizo	respiratory ailments
sarampión, xaltic zahuatl	measles
viruelas, birgoelas, hueyzahuatl, totomanaliztli	smallpox

"The Indians became so enraged by the invulnerability of the Spaniards to epidemic disease that they kneaded infected blood into their masters' bread and secreted corpses in their wells–to little effect" [224, p. 38].

"It is difficult to date precisely when yellow fever first entered the New World. One serious outbreak, acknowledged...as the first, began in Barbados in 1647, reached the Yucatán in 1648, and is recorded for Guadeloupe, Cuba, and Saint Kitts in 1648-1649. Another severe outbreak swept the northeast coast of Brazil from 1686 to 1694. Yellow fever reached Boston in 1693, having been brought there by the British fleet returning from Barbados [229, pp. 227, 228].

Names of Tuberculosis	
consumption	[226, p. 5]
gibbus, Pott's disease	[226, p. 4]
hectic fever	[226, p. 71]
king's evil	[226, p. 7]
lunger	[226, p. 3]
phthisis	[226, p. 5]
poitrinaire	[226, p. 3]
scrofula	[226, p. 4]
tabes	[226, p. 72]

[a] [229, pp. 243-245], [224]

Turing Machines

A Turing machine can be used to describe a variety of processes already described in emergent computation. Obviously, all regular automata, pushdown automata, and linear-bounded automata may be expressed in terms of Turing machines, thus previous papers based upon Chomsky type 3, type 2, and type 1 grammars or languages can be expressed in terms of Turing machines. Here we will describe some approaches in emergent computation that explicitly refer to Turing machine constructions.

The first paper of interest is "Molecular Algorithms" by W.R. Stahl and H. E. Goheen [63, pp. 266-287]. This paper will be summarized.

Five different Turing machines are constructed to show possible applications in emergent computation. The first Turing machine is intended to show how a Turing machine can simulate enzymes that operate on input strings of helical polypeptides and nucleic acids with bases that repeat periodically every four residues. Turing machines can be used to simulate DNA replication, enzyme action, lytic action, etc. It is claimed that perhaps with 30 to 50 such automata, a primitive cell could be simulated.

The first Turing machine has a string on its tape of the form:

$\alpha\beta\gamma\delta\varepsilon$

New residue added here, as a function $f(\alpha, \delta)$ (every symbol, and another symbol three places to the right)

The residue alphabet consists of symbols "a" and "b". The function f is defined as follows:

$$f\left(\alpha, \beta\right) = \begin{cases} a, & \beta = \alpha \\ b, & \beta \neq \alpha \end{cases}$$

Thus we are given a tape described as follows:

$q_1 \underset{\wedge}{\omega}$ where string ω is of length 4 or more, and the string is framed by "h", and "ϕ" indicates an empty tape position (blank on tape).

If $\omega = \phi\phi\phi h\, a\, b\, a\, b\, a\, b\, b\, h\, \phi\phi\phi$ initially, then the Turing machine result is:
$\phi\phi\phi h\, a\, b\, a\, b\, a\, b\, b\, h\, \phi\phi\phi\, a\, a\, b\, a\, a\, a\, b\, b\, b\, b\, a\, h$

The Turing machine is of the 5-tuple type where R means a right move, L means a left move, and P signifies a write with no motion of the read-write head on the tape.

Thus:

$$\delta(q_A,\ x) = q_B R\ y \quad \text{or} \quad \langle q_A,\ x,\ q_B,\ R,\ y \rangle$$

meaning that reading tape symbol "x" in state q_A, the Turing machine moves to state q_B, writes "y" on the tape, and moves right.

$$\delta(q_A,\ x) = q_B L\ y \quad \text{or} \quad \langle q_A,\ x,\ q_B,\ L,\ y \rangle$$

meaning that reading tape symbol "x" in state q_A, the Turing machine moves to state q_B, writes "y" on the tape, and moves left.

$$\delta(q_A,\ x) = q_B P\ y \quad \text{or} \quad \langle q_A,\ x,\ q_B,\ P,\ y \rangle$$

meaning that reading tape symbol "x" in state q_A, the Turing machine moves to state q_B, writes "y" on the tape, and doesn't move.

The following Turing machine accomplishes this copy operation.

	h	a	b	ϕ
q_1	$q_2 Rh$	....	....	$q_1 R\phi$
q_2	$q_3 Lh$	$q_2 Ra$	$q_2 Rb$	....
q_3	$q_3 Ph$	$q_4 La$	$q_7 Lb$	....
q_4	$q_4 Lb$	$q_5 La$	$q_5 Lb$	$q_2 Rh$
q_5	$q_5 La$	$q_6 La$	$q_6 Lb$	$q_2 Rh$
q_6	$q_6 Lb$	$q_{10} Ra$	$q_{11} Rb$	$q_2 Rh$
q_7	$q_7 La$	$q_8 La$	$q_8 Lb$	$q_2 Rh$
q_8	$q_8 Lb$	$q_9 La$	$q_9 Lb$	$q_2 Rh$
q_9	$q_9 Lb$	$q_{11} Ra$	$q_{10} Rb$	$q_2 Rh$
q_{10}	$q_{10} Ra$	$q_{10} Ra$	$q_{10} Rb$	$q_3 Lh$
q_{11}	$q_{11} Rb$	$q_{11} Ra$	$q_{11} Rb$	$q_3 Lh$

Sub-Turing Machines

This copy or replication can easily be expanded to a four-letter alphabet for DNA (with small modification, this same Turing machine can be used in an RNA environment, and the number of bases can be increased forlarger alphabets). For example, let us define three sub-Turing machines as follows, then put them together to obtain a single Turing machine:

T_1 determines if input string ω has any A, C, G, T

T_2 restores input string ω

T_3 converts A to a, C to c, G to g, T to t

The final Turing machine:

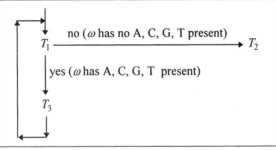

Figure 7.1 Using three sub-Turing machines

Note that now we shall use "Δ" to symbolize a blank on the tape.

T_1	Δ	A	C	G	T	a	c	g	t
q_1	q_2R								
q_2	q_4L	q_3L	q_3L	q_3L	q_3L	q_2R	q_2R	q_2R	q_2R
q_3	h_{yes}	q_3L	q_3L	q_3L	q_3L	q_3L	q_3L	q_3L	q_3L
q_4	h_{no}	q_4L	q_4L	q_4L	q_4L	q_4L	q_4L	q_4L	q_4L

T_2	Δ	A	C	G	T	a	c	g	t
r_1	r_2R								
r_2	r_3L	r_2R	r_2R	r_2R	r_2R	r_2R	r_2R	r_2R	r_2R
r_3	h_{OK}					r_4A	r_4C	r_4G	r_4T
r_4		r_3L	r_3L	r_3L	r_3L				

T_3	Δ	A	C	G	T	a	c	g	t
s_1	$s_2 R$								
s_2	$s_3 L$	$s_2 R$	$s_2 R$	$s_2 R$	$s_2 R$	$s_2 R$	$s_2 R$	$s_2 R$	$s_2 R$
s_3	h_{done}	$s_4 a$	$s_7 c$	$s_{10} g$	$s_{13} t$	$s_3 L$	$s_3 L$	$s_3 L$	$s_3 L$
s_4	$s_5 R$	$s_4 R$	$s_4 R$	$s_4 R$	$s_4 R$	$s_4 R$	$s_4 R$	$s_4 R$	$s_4 R$
s_5	$s_6 T$	$s_5 R$	$s_5 R$	$s_5 R$	$s_5 R$	$s_5 R$	$s_5 R$	$s_5 R$	$s_5 R$
s_6	$s_3 L$	$s_6 L$	$s_6 L$	$s_6 L$	$s_6 L$	$s_6 L$	$s_6 L$	$s_6 L$	$s_6 L$
s_7	$s_8 R$	$s_7 R$	$s_7 R$	$s_7 R$	$s_7 R$	$s_7 R$	$s_7 R$	$s_7 R$	$s_7 R$
s_8	$s_9 G$	$s_8 R$	$s_8 R$	$s_8 R$	$s_8 R$	$s_8 R$	$s_8 R$	$s_8 R$	$s_8 R$
s_9	$s_3 L$	$s_9 L$	$s_9 L$	$s_9 L$	$s_9 L$	$s_9 L$	$s_9 L$	$s_9 L$	$s_9 L$
s_{10}	$s_{11} R$	$s_{10} R$	$s_{10} R$	$s_{10} R$	$s_{10} R$	$s_{10} R$	$s_{10} R$	$s_{10} R$	$s_{10} R$
s_{11}	$s_{12} C$	$s_{11} R$	$s_{11} R$	$s_{11} R$	$s_{11} R$	$s_{11} R$	$s_{11} R$	$s_{11} R$	$s_{11} R$
s_{12}	$s_3 L$	$s_{12} L$	$s_{12} L$	$s_{12} L$	$s_{12} L$	$s_{12} L$	$s_{12} L$	$s_{12} L$	$s_{12} L$
s_{13}	$s_{14} R$	$s_{13} R$	$s_{13} R$	$s_{13} R$	$s_{13} R$	$s_{13} R$	$s_{13} R$	$s_{13} R$	$s_{13} R$
s_{14}	$s_{15} A$	$s_{14} R$	$s_{14} R$	$s_{14} R$	$s_{14} R$	$s_{14} R$	$s_{14} R$	$s_{14} R$	$s_{14} R$
s_{15}	$s_3 L$	$s_{15} L$	$s_{15} L$	$s_{15} L$	$s_{15} L$	$s_{15} L$	$s_{15} L$	$s_{15} L$	$s_{15} L$

Putting these three Turing machines together, we obtain the following Turing machine.

	Δ	A	C	G	T	a	c	g	t
q_1	q_2R								
q_2	q_4L	q_3L	q_3L	q_3L	q_3L	q_2R	q_2R	q_2R	q_2R
q_3	$s_1\Delta$	q_3L	q_3L	q_3L	q_3L	q_3L	q_3L	q_3L	q_3L
q_4	$r_1\Delta$	q_4L	q_4L	q_4L	q_4L	q_4L	q_4L	q_4L	q_4L
r_1	r_2R								
r_2	r_3L	r_2R	r_2R	r_2R	r_2R	r_2R	r_2R	r_2R	r_2R
r_3	h_{OK}					r_4A	r_4C	r_4G	r_4T
r_4		r_3L	r_3L	r_3L	r_3L				
s_1	s_2R								
s_2	s_3L	s_2R	s_2R	s_2R	s_2R	s_2R	s_2R	s_2R	s_2R
s_3	$q_1\Delta$	s_4a	s_7c	$s_{10}g$	$s_{13}t$	s_3L	s_3L	s_3L	s_3L
s_4	s_5R	s_4R	s_4R	s_4R	s_4R	s_4R	s_4R	s_4R	s_4R
s_5	s_6T	s_5R	s_5R	s_5R	s_5R	s_5R	s_5R	s_5R	s_5R
s_6	s_3L	s_6L	s_6L	s_6L	s_6L	s_6L	s_6L	s_6L	s_6L
s_7	s_8R	s_7R	s_7R	s_7R	s_7R	s_7R	s_7R	s_7R	s_7R
s_8	s_9G	s_8R	s_8R	s_8R	s_8R	s_8R	s_8R	s_8R	s_8R
s_9	s_3L	s_9L	s_9L	s_9L	s_9L	s_9L	s_9L	s_9L	s_9L
s_{10}	$s_{11}R$	$s_{10}R$	$s_{10}R$	$s_{10}R$	$s_{10}R$	$s_{10}R$	$s_{10}R$	$s_{10}R$	$s_{10}R$
s_{11}	$s_{12}C$	$s_{11}R$	$s_{11}R$	$s_{11}R$	$s_{11}R$	$s_{11}R$	$s_{11}R$	$s_{11}R$	$s_{11}R$
s_{12}	s_3L	$s_{12}L$	$s_{12}L$	$s_{12}L$	$s_{12}L$	$s_{12}L$	$s_{12}L$	$s_{12}L$	$s_{12}L$
s_{13}	$s_{14}R$	$s_{13}R$	$s_{13}R$	$s_{13}R$	$s_{13}R$	$s_{13}R$	$s_{13}R$	$s_{13}R$	$s_{13}R$
s_{14}	$s_{15}A$	$s_{14}R$	$s_{14}R$	$s_{14}R$	$s_{14}R$	$s_{14}R$	$s_{14}R$	$s_{14}R$	$s_{14}R$
s_{15}	s_3L	$s_{15}L$	$s_{15}L$	$s_{15}L$	$s_{15}L$	$s_{15}L$	$s_{15}L$	$s_{15}L$	$s_{15}L$

Similar to the discussions in Chapters 4 and 5 we could modify this language by supporting non-standard bases, non-standard chelated bases, totally artificial bases, abasic pairs, etc. as discussed in Chapter 3. We would have to add the following to any such Turing machine: isoC, isoG, ic, ig, K, Π, κ, π, E, e, H, h, Dipic, Py, di, py, 5MICS, 5m, MICS, m, P, $\varnothing$, p, $\varnothing$, etc. as tape symbols, and symbols that can be written also–really, this is only a detail!

$$
\begin{array}{llllllllllll}
q_1 & \Delta & A & C & T & G & \Delta & \Delta & \Delta & \Delta & \Delta & \Delta \\
\Delta & q_2 & A & C & T & G & \Delta & \Delta & \Delta & \Delta & \Delta & \Delta \\
q_3 & \Delta & A & C & T & G & \Delta & \Delta & \Delta & \Delta & \Delta & \Delta \\
s_1 & \Delta & A & C & T & G & \Delta & \Delta & \Delta & \Delta & \Delta & \Delta \\
\Delta & s_2 & A & C & T & G & \Delta & \Delta & \Delta & \Delta & \Delta & \Delta \\
\Delta & A & s_2 & C & T & G & \Delta & \Delta & \Delta & \Delta & \Delta & \Delta \\
\end{array}
$$

$\bullet \quad \bullet \quad \bullet$

$$
\begin{array}{lllllllllll}
\Delta & A & C & T & G & s_2 & \Delta & \Delta & \Delta & \Delta & \Delta \\
\Delta & A & C & T & s_3 & G & \Delta & \Delta & \Delta & \Delta & \Delta \\
\Delta & A & C & T & s_{10} & g & \Delta & \Delta & \Delta & \Delta & \Delta \\
\Delta & A & C & T & g & s_{10} & \Delta & \Delta & \Delta & \Delta & \Delta \\
\Delta & A & C & T & g & \Delta & s_{11} & \Delta & \Delta & \Delta & \Delta \\
\Delta & A & C & T & g & \Delta & s_{12} & C & \Delta & \Delta & \Delta \\
\Delta & A & C & T & g & s_{12} & \Delta & C & \Delta & \Delta & \Delta \\
\Delta & A & C & T & s_3 & g & \Delta & C & \Delta & \Delta & \Delta \\
\Delta & A & C & s_3 & T & g & \Delta & C & \Delta & \Delta & \Delta \\
\Delta & A & C & s_{13} & t & g & \Delta & C & \Delta & \Delta & \Delta \\
\end{array}
$$

$\bullet \quad \bullet \quad \bullet$

$$
\begin{array}{lllllllllll}
\Delta & A & C & t & g & s_{13} & \Delta & C & \Delta & \Delta & \Delta \\
\Delta & A & C & t & g & \Delta & s_{14} & C & \Delta & \Delta & \Delta \\
\Delta & A & C & t & g & \Delta & C & s_{14} & \Delta & \Delta & \Delta \\
\Delta & A & C & t & g & \Delta & C & s_{15} & A & \Delta & \Delta & \Delta \\
\end{array}
$$

$\bullet \quad \bullet \quad \bullet$

$$
\begin{array}{lllllllllll}
\Delta & A & C & t & g & s_{15} & \Delta & C & A & \Delta & \Delta & \Delta \\
\Delta & A & C & t & s_3 & g & \Delta & C & A & \Delta & \Delta & \Delta \\
\Delta & A & C & s_3 & t & g & \Delta & C & A & \Delta & \Delta & \Delta \\
\Delta & A & s_3 & C & t & g & \Delta & C & A & \Delta & \Delta & \Delta \\
\Delta & A & s_7 & c & t & g & \Delta & C & A & \Delta & \Delta & \Delta \\
\end{array}
$$

(continued)

$\Delta \quad A \quad c \quad t \quad g \quad s_7 \quad \Delta \quad C \quad A \quad \Delta \quad \Delta \quad \Delta$
$\Delta \quad A \quad c \quad t \quad g \quad \Delta \quad s_8 \quad C \quad A \quad \Delta \quad \Delta \quad \Delta$

$\bullet \quad \bullet \quad \bullet$

$\Delta \quad A \quad c \quad t \quad g \quad \Delta \quad C \quad A \quad s_8 \quad \Delta \quad \Delta \quad \Delta$
$\Delta \quad A \quad c \quad t \quad g \quad \Delta \quad C \quad A \quad s_9 \quad \Delta \quad \Delta \quad \Delta$

$\bullet \quad \bullet \quad \bullet$

$\Delta \quad A \quad c \quad t \quad g \quad s_9 \quad \Delta \quad C \quad A \quad G \quad \Delta \quad \Delta$
$\Delta \quad A \quad c \quad t \quad s_3 \quad g \quad \Delta \quad C \quad A \quad G \quad \Delta \quad \Delta$

$\bullet \quad \bullet \quad \bullet$

$\Delta \quad s_3 \quad A \quad c \quad t \quad g \quad \Delta \quad C \quad A \quad G \quad \Delta \quad \Delta$
$\Delta \quad s_4 \quad a \quad c \quad t \quad g \quad \Delta \quad C \quad A \quad G \quad \Delta \quad \Delta$

$\bullet \quad \bullet \quad \bullet$

$\Delta \quad a \quad c \quad t \quad g \quad s_4 \quad \Delta \quad C \quad A \quad G \quad \Delta \quad \Delta$
$\Delta \quad a \quad c \quad t \quad g \quad \Delta \quad s_5 \quad C \quad A \quad G \quad \Delta \quad \Delta$

$\bullet \quad \bullet \quad \bullet$

$\Delta \quad a \quad c \quad t \quad g \quad \Delta \quad C \quad A \quad G \quad s_5 \quad \Delta \quad \Delta$
$\Delta \quad a \quad c \quad t \quad g \quad \Delta \quad C \quad A \quad G \quad s_6 \quad T \quad \Delta$

$\bullet \quad \bullet \quad \bullet$

$\Delta \quad a \quad c \quad t \quad g \quad s_6 \quad \Delta \quad C \quad A \quad G \quad T \quad \Delta$
$\Delta \quad a \quad c \quad t \quad s_3 \quad g \quad \Delta \quad C \quad A \quad G \quad T \quad \Delta$

$\bullet \quad \bullet \quad \bullet$

$s_3 \quad \Delta \quad a \quad c \quad t \quad g \quad \Delta \quad C \quad A \quad G \quad T \quad \Delta$
$q_1 \quad \Delta \quad a \quad c \quad t \quad g \quad \Delta \quad C \quad A \quad G \quad T \quad \Delta$
$\Delta \quad q_2 \quad a \quad c \quad t \quad g \quad \Delta \quad C \quad A \quad G \quad T \quad \Delta$

$\bullet \quad \bullet \quad \bullet$

$\Delta \quad a \quad c \quad t \quad g \quad q_2 \quad \Delta \quad C \quad A \quad G \quad T \quad \Delta$
$\Delta \quad a \quad c \quad t \quad q_4 \quad g \quad \Delta \quad C \quad A \quad G \quad T \quad \Delta$
$q_4 \quad \Delta \quad a \quad c \quad t \quad g \quad \Delta \quad C \quad A \quad G \quad T \quad \Delta$
$r_1 \quad \Delta \quad a \quad c \quad t \quad g \quad \Delta \quad C \quad A \quad G \quad T \quad \Delta$
$\Delta \quad r_2 \quad a \quad c \quad t \quad g \quad \Delta \quad C \quad A \quad G \quad T \quad \Delta$

$\bullet \quad \bullet \quad \bullet$

(continued)

Δ	a	c	t	g	r_2	Δ	C	A	G	T	Δ
Δ	a	c		r_3	g	Δ	C	A	G	T	Δ
Δ	a	c		r_4	G	Δ	C	A	G	T	Δ
Δ	a	c	r_3	t	G	Δ	C	A	G	T	Δ
Δ	a	c	r_4	T	G	Δ	C	A	G	T	Δ
Δ	a	r_3	c	T	G	Δ	C	A	G	T	Δ
Δ	a	r_4	C	T	G	Δ	C	A	G	T	Δ
Δ	r_3	a	C	T	G	Δ	C	A	G	T	Δ
Δ	r_4	A	C	T	G	Δ	C	A	G	T	Δ
r_3	Δ	A	C	T	G	Δ	C	A	G	T	Δ
h_{OK}											

Thus we obtain, in linear form, $A\ C\ T\ G\ \ C\ A\ G\ T$ and if we "bend" this around, we have the typical base sequence found on a DNA double-stranded helix.

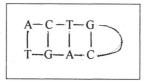

Figure 7.2

Turing Machines that Simulate Enzyme Action

The use of Turing machines to model enzyme action is illustrated by an example as follows:

$$ac + a \;\rightarrow\; aca \qquad \text{whenever enzyme ``}bb\text{'' is present, using energy ``}e\text{''}$$

energy "e" is consumed, converting it into ϕ
enzyme "bb" is not consumed
"$*$" is used to separate words on the tape

Note that energy "e" could be converted into energy "d" to model the conversion of ATP to ADP, for example.

The Turing machine is given on the next page.

An example of such an enzymatic transformation is as follows.

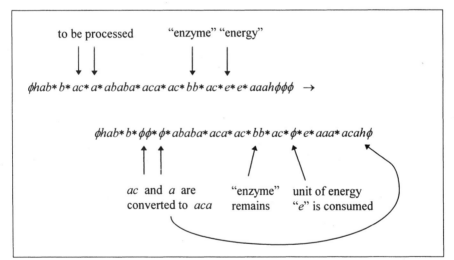

Figure 7.3 Enzyme activity simulated by a Turing machine

The corresponding Turing machine is on the next page.

	h	$*$	ϕ	a	b	c	e	x
q_1	q_2Rh		$q_1R\phi$	$q_{11}Lx$				
q_2	q_2Ph	q_2R*	$q_2R\phi$	q_2Ra	q_2Rb	q_2Rc	q_3Le	
q_3	q_4Rh	q_3L*	$q_3L\phi$	q_3La	q_3Lb	q_3Lc		
q_4		q_4P*	$q_8R\phi$	q_8Ra	q_5Rb	q_8Rc	q_8Re	
q_5		q_4R*		q_8Ra	q_6Rb	q_8Rc		
q_6	q_7Lh	q_7L*		q_8Ra	q_8Rb	q_8Rc		
q_7	q_9Rh	q_7L*	$q_7L\phi$	q_7La	q_7Lb	q_7Lc	q_7Le	
q_8	q_8Ph	q_4R*	$q_8R\phi$	q_8Ra	q_8Rb	q_8Rc	q_8Pe	
q_9			$q_{12}R\phi$	$q_{10}Ra$	$q_{12}Rb$	$q_{12}Rc$	$q_{12}Re$	
q_{10}	q_1Lh	q_1L*		$q_{12}Ra$	$q_{12}Rb$	$q_{12}Rc$		
q_{11}	$q_{13}Rh$	$q_{11}L*$	$q_{11}L\phi$	$q_{11}La$	$q_{11}Lb$	$q_{11}Lc$	$q_{11}Le$	
q_{12}	$q_{12}Ph$	q_9R*	$q_{12}R\phi$	$q_{12}Ra$	$q_{12}Rb$	$q_{12}Rc$		
q_{13}			$q_{17}R\phi$	$q_{14}Ra$	$q_{17}Rb$	$q_{17}Rc$	$q_{17}Re$	$q_{17}Rx$
q_{14}	$q_{18}Lh$	$q_{13}R*$		$q_{17}Ra$	$q_{17}Rb$	$q_{15}Rc$		
q_{15}	$q_{16}Lh$	$q_{16}L*$		$q_{17}Ra$	$q_{17}Rb$	$q_{17}Rc$		
q_{16}	$q_{20}Rh$		$q_{19}L*$	$q_{16}L\phi$		$q_{16}L\phi$		
q_{17}	$q_{18}Lh$	$q_{13}R*$	$q_{17}R\phi$	$q_{17}Ra$	$q_{17}Rb$	$q_{17}Rc$		
q_{18}	$q_{18}Ph$	$q_{18}L*$	$q_{18}L\phi$	$q_{18}La$	$q_{18}Lb$	$q_{18}Lc$	$q_{18}Le$	$q_{18}La$
q_{19}	$q_{20}Rh$	$q_{19}L*$	$q_{19}L\phi$	$q_{19}La$	$q_{19}Lb$	$q_{19}Lc$	$q_{19}Le$	$q_{19}Lx$
q_{20}		$q_{20}R*$	$q_{20}R\phi$	$q_{20}Ra$	$q_{20}Rb$	$q_{20}Rc$	$q_{21}R\phi$	$q_{20}R\phi$
q_{21}	$q_{22}R*$	$q_{21}R*$	$q_{21}R\phi$	$q_{21}Ra$	$q_{21}Rb$	$q_{21}Rc$	$q_{21}Re$	$q_{21}R\phi$
q_{22}			$q_{23}Ra$					
q_{23}			$q_{24}Rc$					
q_{24}			$q_{25}Ra$					
q_{25}			$q_{26}Lh$					
q_{26}	q_2Rh	$q_{26}L*$	$q_{26}L\phi$	$q_{26}La$	$q_{26}Lb$	$q_{26}Lc$	$q_{26}Le$	

Turing Machines and DNA Replication

In the next Turing machine, Goheen and Stahl [63] show how DNA can be replicated, although I have constructed a more complicated replication Turing machine that preserves Watson-Crick complements. The Turing machine will do the following replication.

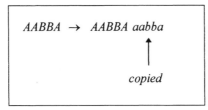

Figure 7.4

This will be done in detail, as follows.

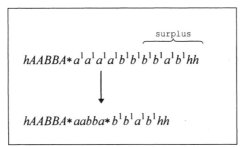

Figure 7.5

The corresponding Turing machine is on the next page.

	h	A	B	ϕ	$*$	a^1	b^1	$\#$	a	b
q_1	q_2Rh	$q_3R\phi$	$q_4R\phi$		q_2R*	$q_{13}R*$	$q_{14}R*$		q_2Ra	q_2Rb
q_2	$q_{15}R\#$	q_3RA	q_3RB		q_3R*	q_5Pa	$q_7R\#$		q_3Ra	q_3Rb
q_3	$q_{12}Lh$	q_4RA	q_4RB		q_4R*	$q_8R\#$	q_6Pb		q_4Ra	q_4Rb
q_4	$q_{11}Lh$	q_5LA	q_5LB	q_2RA	q_5L*				q_5La	q_5Lb
q_5		q_6LA	q_6LB	q_2RB	q_6L*				q_6La	q_6Lb
q_6										
q_7	$q_{12}Lh$					q_9Lb^1	q_7Rb^1		$q_{12}Pa^1$	$q_{11}Pb^1$
q_8	$q_{11}Lh$					q_8Ra^1	$q_{10}La^1$			
q_9							q_9Lb^1	q_5Pa		
q_{10}						$q_{10}La^1$		q_6Pb		
q_{11}	$q_{11}Ph$	$q_{11}LA$	$q_{11}LB$		$q_{11}L*$	$q_{11}La^1$	$q_{11}Lb^1$	$q_{11}La^1$	$q_{11}La^1$	$q_{11}Lb^1$
q_{12}	$q_{12}Ph$	$q_{12}LA$	$q_{12}LB$		$q_{12}L*$	$q_{11}La^1$	$q_{12}Lb^1$	$q_{12}Lb^1$	$q_{12}La^1$	$q_{12}Lb^1$
q_{13}	$q_{15}Ra^1$					$q_{13}Ra^1$	$q_{13}Rb^1$			
q_{14}	$q_{15}Rb^1$					$q_{14}Ra^1$	$q_{14}Rb^1$			
q_{15}	$q_{15}Ph$									

Turing Machines and Lysis

A Turing machine can not only replicate or copy, but can also split polypeptide chains. An example is provided where a b-c bond may be split, causing lysis, but only in specific contexts. The b-c bond may be broken only if there is at least one letter to the left of b and one letter to the right of c. Thus in the case that $\rho, \omega \in \{a, b, c\}^*$.

$$\rho b - c\omega \rightarrow \rho b \qquad c\omega, \qquad b - c \ bond \ is \ split$$

$$\left. \begin{array}{ll} \rho b - c & b - c \ bond \ will \ not \ be \ split \\ b - c\omega & b - c \ bond \ will \ not \ be \ split \\ b - c & b - c \ bond \ will \ not \ be \ split \end{array} \right\}$$

Figure 7.6 Lysis rules

The Turing machine that follows will convert e^1 to e for each b-c bond that is split. The following is an example.

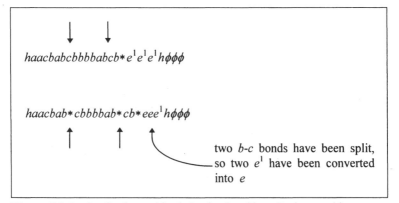

Figure 7.7 Lysis of polypeptide chain simulated by a Turing machine

The corresponding Turing machine is on the next page.

	h	a	b	c	$*$	ϕ	e	e^1	$\#$
q_1	$q_2 Rh$	$q_1 La$	$q_1 Lb$	$q_1 Lc$	$q_1 L*$	$q_1 R\phi$	$q_{11} Pe^1$	$q_1 Pe^1$	
q_2	$q_2 Ph$	$q_2 Ra$	$q_2 Rb$	$q_2 Rc$	$q_2 R*$	$q_9 Lh$	$q_2 Re$	$q_3 Le$	
q_3	$q_4 Rh$	$q_3 La$	$q_3 Lb$	$q_3 Lc$	$q_3 L*$	$q_9 La$	$q_3 Le$		
q_4	$q_1 Lh$	$q_5 Ra$	$q_5 Rb$	$q_5 Rc$	$q_4 R*$	$q_9 Lb$	$q_4 Re$	$q_4 Re^1$	
q_5	$q_1 Lh$	$q_5 Ra$	$q_6 Rb$	$q_5 Rc$	$q_4 R*$	$q_9 Lc$			
q_6	$q_1 Lh$	$q_5 Pa$	$q_5 Pb$	$q_7 R\#$	$q_4 R*$	$q_9 L*$			
q_7	$q_7 Lh$	$q_8 Ra$	$q_8 Rb$	$q_8 Rc$	$q_7 L*$	$q_9 Le$			$q_4 Pc$
q_8	$q_2 R\phi$	$q_8 Ra$	$q_8 Rb$	$q_8 Rc$	$q_8 R*$	$q_9 Le^1$	$q_8 Re$	$q_8 Re^1$	
q_9	$q_1 Lh$	$q_3 R\phi$	$q_4 R\phi$	$q_5 R\phi$	$q_6 R\phi$	$q_9 L\phi$	$q_7 R\phi$	$q_8 R\phi$	$q_{10} R*$
q_{10}	$q_2 Rh$	$q_{10} La$	$q_{10} Lb$	$q_{10} Lc$	$q_{10} L*$	$q_{10} Pc$	$q_{10} Le$	$q_{10} Le^1$	
q_{11}	$q_{11} Ph$	$q_{11} La$	$q_{11} Lb$	$q_{11} Lc$	$q_{11} L*$		$q_{11} Le$	$q_{11} Le^1$	

In Vitro DNA Universal Turing Machines

This book will, at times, focus not upon the interrelationships between automata theory (mathematical linguistics) and aspects of bioinformatics, but will instead focus more upon an actual implementation (in vitro or in vivo). We now discuss attempts (at least on paper, if not in the laboratory) to create Turing machine implementations in a bioinformatics setting. In case there is some doubt that Rothemund has the construction of a "molecular computer," let us quote.

"In order to construct a general molecular computer some Universal model of computation (e.g. a digital computer, neural network, Turing machine, etc.) must be expressed in chemistry" [144, p. 76]. Even more to the point, "In this paper we present a method for encoding the transition table of a Turing machine with oligonucleotides, representing a Turing tape, head position, and state, as a single molecule of DNA, and effecting transitions using restriction enzyme chemistry."..."Using this method a Universal Turing machine (capable of simulating any Turing machine and hence any algorithm) can be constructed" [144, p. 77]. Rothemund points out that there are 67 atoms per A/T pair and 66 atoms per C/G pair (including the deoxyribose/phosphate backbone and two Na^+ ions per base pair), providing 1 bit per 33 atoms for double-stranded DNA. Thus there are 0.33 kg DNA per mole bits, or 0.33 kg DNA per 6.02×10^{23} bits (a lot of memory in very little weight) [144, p. 84].

Using restriction endonucleases, circular plasmid DNA may be cleaved in a controlled fashion, new oligonucleotide material may be inserted, and the plasmid DNA ligated. Similarly, circular plasmid DNA may be cleaved, and then ligated, thereby deleting a section of DNA from the plasmid. Obviously, these two operations may be combined, thus replacing one section of DNA by another in a plasmid DNA. Thus see Figures 7.8 through 7.13.

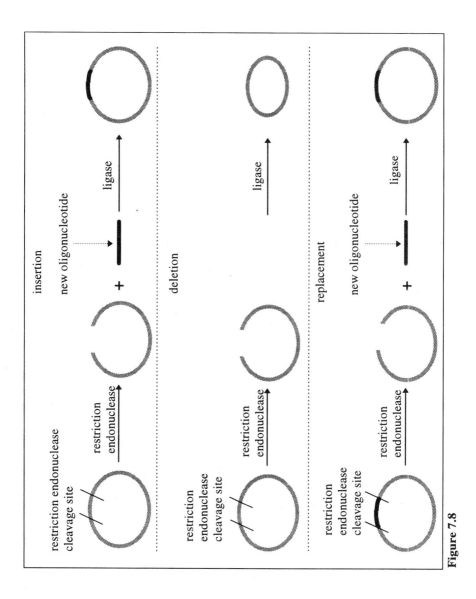

Figure 7.8

Rothemund provides details for a number of Turing machine operations that would be required in a DNA computer.

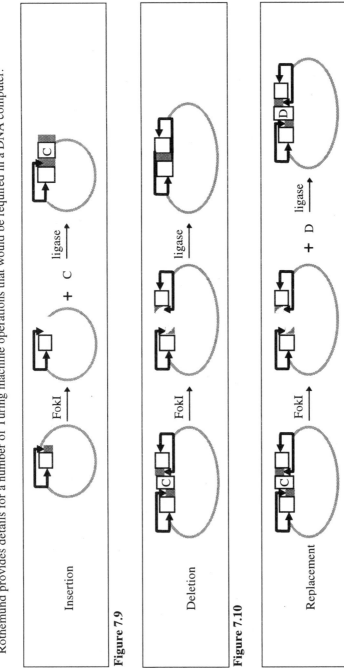

Figure 7.9

Figure 7.10

Figure 7.11

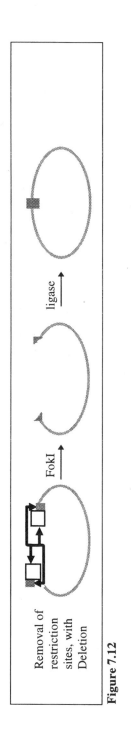

Figure 7.12

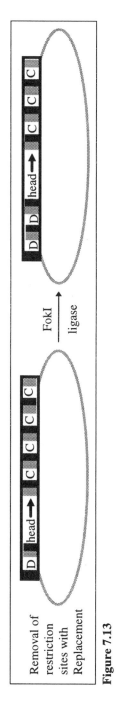

Figure 7.13

With the basic operations discussed, short sequences of oligonucleotides may be added or deleted within circular plasmid DNA. Thus we can encode state information (an oligonucleotide can be used to identify a specific state), and we can position this information. Similarly, a direction (left or right), may be encoded as another oligonucleotide and appended wherever it is required in the plasmid. It turns out that an orientation (as in left or right) is required, and this may be accomplished by using the asymmetric overhangs specific to different "universal" restriction endonucleases. Similarly, data may be encoded as oligonucleotides and placed within the plasmid. Thus an Instantaneous Description (ID) as well known and described by Martin Davis [190], may be recorded within the plasmid. As the Turing machine transition table effectively is a quadruple or quintuple (depending upon how the Turing machine is constructed), all that is required can be accomplished using a plasmid and restriction endonucleases and ligases. For example, the Turing machine tuple might be:

$$\langle q_i, \ T_j, \ q_n, \ T_k, \ L \rangle \qquad \langle q_i, \ T_j, \ q_n, \ T_k, \ R \rangle \qquad \langle q_i, \ T_j, \ q_n, \ --, \ H \rangle$$

where q_i is the current state, reading tape symbol T_j, and as a result moving to state q_n, writing (replacing T_j) with tape symbol T_k, and finally moving left "L" (or right "R", or halting "H").

To accomplish this, Rothemund requires "Coh", "Em", "Inv", "Sta", "Cap", "Res", "R", "L", "X" (symbol to mark excision site) special "symbols." All may be represented by different restriction endonucleases [144, p. 99]. Thus the following restriction endonucleases are utilized [144, p. 104].

Em	is	BbvI
Sta (current)	is	FokI
Sta (next)	is	HgaI
Inv	is	BseRI
Cap	is	BsrDI
X	is	BpmI

These restriction endonucleases are described in Figure 7.14.

The following restriction endonucleases are referred to by Rothemund.

Bacillus brevis	BbvI	5'—G C A G C (N)$_8$ $\downarrow$—3' 3'—C G T C G (N)$_{12}$—5' $\uparrow$
Bacillus pumilus	BpmI	5'—C T G G A G (N)$_{16}$ $\downarrow$—3' 3'—G A C C T C (N)$_{14}$—5' $\uparrow$
Bacillus R	BseRI	5'—G A G G A G (N)$_{10}$ $\downarrow$—3' 3'—C T C C T C (N)$_8$—5' $\uparrow$
Bacillus stearothermophilus	BsrDI	5'—G C A A T G (N)$_2$ $\downarrow$—3' 3'—C G T T A C (N)$_2$—5' $\uparrow$
Flavobacterium okeanokoites	FokI	5'—G G A T G (N)$_9$ $\downarrow$—3' 3'—C C T AC (N)$_{13}$—5' $\uparrow$
Haemophilus gallinarum	HgaI	5'—G A C G C (N)$_5$ $\downarrow$—3' 3'—C T G C G (N)$_{10}$—5' $\uparrow$

Figure 7.14

In Vitro DNA Turing Machine Errors

What might some of the problems be, if we proceed with our universal DNA Turing machine implementation? Rothemund lists some problems.

Problem 1: If we use a unique restriction endonuclease for each (unique) state/symbol table transition, then many restriction endonucleases would be required and many chemical reactions.

Problem 2: Chemical reactions other than the desired ones might take place, such as:

A: The creation of dimers, trimers, and higher "mers".

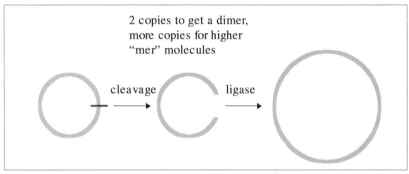

Figure 7.15

B: Multiple repeated inserts of the same oligonucleotide sequence (concatamers).

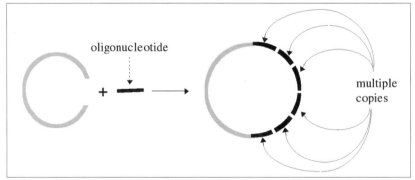

Figure 7.16

C: Same as in case B, except possible reversed order (polarity) of inserted sequences of oligonucleotides.

D: Other possibilities, which are examined in Rothemund.

Many of these problems are caused by palindromic staggered-ended restriction nucleases, which in addition, may have unique staggered (sticky) ends. Thus using non-palindromic (asymmetric) restriction endonucleases alleviates this problem. An example of such a restriction endonuclease is provided by FokI. In addition, FokI allows many different sticky ends, by varying the $13 - 9 = 4$ different nucleotide bases! If greater variation is required, Szybalski ([172, pp. 169-173] and [173, pp. 13-26]) has provided "universal restriction endonucleases."

While we do not describe all the possible maneuvers required, Rothemund can be consulted for details. However, significant problems arise when one considers errors. Rothemund discusses errors and strategies to deal with these errors [144, pp. 107-110].

A. Failed ligations. The defective linear plasmid may be "chewed up" using exonucleases III and S1.

B. Incorrect ligations. Ligation of mismatched sticky (staggered) ends. Such errors may be detected using nuclease S1, then exonuclease III to remove these errors.

C. Failed restrictions. Biotin or other molecules may be used to mark defective "tapes" at head, Cap, or X sites. After each restriction, all DNA tapes still marked may be removed. Rothemund mentions different molecules that could easily be used as markers.

D. Incorrect restrictions. Errors due to cleavage at the wrong locations rarely occur with restriction endonucleases. In most cases, such errors take place under non-standard reaction conditions such as temperature, pressure, concentration, etc. These factors can be controlled using standard buffers, standard temperatures, etc.

E. Dimerization. To some degree, dimerization is caused by steric stresses, and this can be controlled using CAP proteins to reduce steric stresses.

It is very interesting that in vitro implementation of a DNA Turing machine is possible. In such a situation, the energy required to run such a "computer" is not supplied by external electrical power, but rather by converting ATP to ADP or AMP thereby releasing energy [144, p. 112]! Recall that we have encountered in vitro and in vivo applications already in Chapter 3 as well as in Chapter 4.

Besides the Rothemund paper described in the Lipton citation [144], is the article "DNA Computers in vitrio and vivo" by W. Smith in the same citation [144]. This paper is very similar to the previous one by Rothemund, but notes that such a DNA Turing machine is non-deterministic. Smith attempts to increase efficiency, but feels that DNA Turing machines are not practical.

Smith finds additional sources of error and possible solutions for these errors, such as the following.

Palindromic substrings that may fold into hairpins which can stall polymerases.

Modified "DeBruijn" sequences that avoid mismatches.

The creation of "nicks" due to hydrolysis (usually due to contamination). These problems can be reduced by using sealed vessels in a nitrogen atmosphere, and even DNA repair enzymes.

Smith points out that with minor modification, his proposed DNA Turing machine head can move both left and right. In addition, basic operations might include:

A "bootstrap" loader analogy that initializes the tape.

Recognition of halt states using anti-sense DNA bound to dye as an indicator.

Smith points out that while such DNA computers are not too useful, he nevertheless also simultaneously takes the opposite viewpoint. Smith points out that mRNA in T. Brucei trypanosome is "edited," and that in fact it is possible that mRNA acts as an in vivo Turing machine.

Chapter 8: Splicing Systems, H Systems

<table>
<tr><td>

127. In the course of the fifth year the pestilence began, O my children. First there was a cough, then the blood was corrupted, and the urine became yellow. The number of deaths at this time was terrible. The chief Vakaki Ahmak died, and we ourselves were plunged in great darkness and great grief, our fathers and ancestors having contracted the plague, O my children.

On the day 1 Ah there were one cycle and 5 years from the Revolt, and the pestilence spread.

128. In this year the pestilence spread, and then died our ancestor Diego Juan. On day 5 Ah war was carried to Panatacat by our ancestor, and then began the spread of the pestilence. Truly the number of deaths among the people was terrible, nor did the people escape from the pestilence.

129. Forty were seized with the sickness; then died our father and ancestor; on the day 14 Camey died the king Hunyg, your grandfather.

130. But two days afterward died our father, the Counselor Balam, one of the ancients, O my children. The ancients and the fathers died alike, and the stench was such that men died of it alone. Then perished our fathers and ancestors. Half the people threw themselves into the ravines, and the dogs and foxes lived on the bodies of men...

</td><td>

127. Chupam 4-a voo huna, vae ok ixtiquer yauabil, yxnu4-ahol, nabey xyabix ohb, ratzam xyavavabix chi4-a qui4- , ε ana chuluh, kitzih tixibin chi camic xi4-go oher. Haok xcam ahauh Vakaki Ahmak, xe 4-a hala chic ma tipe nima ε ekum, nima aε a pa qui vi ka tata ka mama pa ka vi 4-a, yxnu4-ahol, ok xyabix 4-hac.

Chi hun Ah xel humay voo yuhuh, ok xyabix 4-hac.

128. Vae chupam huna xyauabix vi 4-hac, ha ok xe4- iz ci camic ka tata ka mama Diego Juan; chi voo Ah 4-axoc chi vi labal Panatacat, cuma ka mama, ha4-a ok xtiquer yavabil 4-hac. Kitzih tixibin chi camic xpe pa ru vi vinak, mani yabim viri quere ri x4- hol vinak.

129. Xcavinak ok xtiquer yauabil, tok xecam ka tata ka mama, chi cablahuh Camey xcam ahauh Hunyε yxiquin mama.

130. Xa4-a ru cabih xcam chic ka tata rahpop Achi Balam ri y mama, yxnu4- ahol; xa 4-a hunam xecam y mama ru4- in tata ki tan ti chuvin, ti 4-ayin vinak chi camic. Tok xecam ka tata ka mama, xax be tzak chi el 4- hakap vinak chi civan, xa 4-,ij xa4-uch, xtiochic vinak; tixibin chi camic xecamiçan ymama, herach camic ru 4-ahol ahuah ru4- in ru chaε ru nimal ...

</td></tr>
</table>

[a] Using the alphabet of missionary Francisco de la Parra (d. 1560), [233, pp. 49-51, 171].

Splicing Systems and Languages

Splicing systems and splicing languages are defined as a formalization of the products of restriction enzymes. Head views the DNA double helix not as two strings, but rather as a single string over couples chosen from A, C, G, and T. We will consider circular splicing systems soon.

$$\text{Explicitly:} \quad D = \left\{ \begin{pmatrix} A \\ T \end{pmatrix}, \begin{pmatrix} T \\ A \end{pmatrix}, \begin{pmatrix} C \\ G \end{pmatrix}, \begin{pmatrix} G \\ C \end{pmatrix} \right\}$$

Note that in this view, the alphabet D, as well as other considerations, assumes:
  no mismatches
  no triple helices, quadruple helices, etc.
  no abasic pairs
  no non-naturally occurring bases in general
  linear DNA (no circular occurring DNA or supercoiled DNA)
  no transposon activity (transposons will be discussed)
  no supercoiling

Thus reviewing Chapter 3, we exclude from our alphabet such objects as:

$\begin{pmatrix} A \\ C \end{pmatrix}, \begin{pmatrix} G \\ G \end{pmatrix}$, etc.   no mismatches

$\begin{pmatrix} A \\ T \\ A \end{pmatrix}, \begin{pmatrix} C \\ G \\ C \end{pmatrix}$, etc.   no triple-stranded DNA

$\begin{pmatrix} P \\ \phi \end{pmatrix}, \begin{pmatrix} \phi \\ P \end{pmatrix}$, etc.   no abasic pairs

$\begin{pmatrix} \text{iso–}C \\ \text{iso–}G \end{pmatrix}, \begin{pmatrix} \text{iso–}G \\ \text{iso–}C \end{pmatrix}, \begin{pmatrix} H \\ H \end{pmatrix}$, etc.   no non-naturally occurring bases, chelates

no quadruple DNA helicies (G-quartets)

no circular DNA, no supercoiled DNA

To continue with our formal definition of splicing systems and splicing languages; unary and binary operations exist over the alphabet D, as follows.

Concatenation, thus $x, y \in D^*$, then $xy \in D^*$

(Note: $\begin{pmatrix} \lambda \\ \lambda \end{pmatrix} \in D^*$, and no molecule corresponds to this.)

length: $D^* \to D^*$; thus for $x = a_1 \ldots a_n$, then $length(x) = n$, and $length(\lambda) = 0$

Involution,   thus $x \in D^*$,   then $f : D^* \to D^*$,

$$\text{where: } f^2(x) = x, \text{ and } f(xy) = f(y) f(x)$$

$$f\left[\begin{pmatrix} A \\ T \end{pmatrix}\right] = \begin{pmatrix} T \\ A \end{pmatrix}, \quad f\left[\begin{pmatrix} T \\ A \end{pmatrix}\right] = \begin{pmatrix} A \\ T \end{pmatrix}, \quad f\left[\begin{pmatrix} C \\ G \end{pmatrix}\right] = \begin{pmatrix} G \\ C \end{pmatrix}, \quad f\left[\begin{pmatrix} G \\ C \end{pmatrix}\right] = \begin{pmatrix} C \\ G \end{pmatrix}$$

thus as an example, let $x \in D^*$, where $x = \begin{pmatrix} G \\ C \end{pmatrix}\begin{pmatrix} A \\ T \end{pmatrix}\begin{pmatrix} C \\ G \end{pmatrix}\begin{pmatrix} A \\ T \end{pmatrix}$, then

$$f[x] = f\left[\begin{pmatrix} G \\ C \end{pmatrix}\begin{pmatrix} A \\ T \end{pmatrix}\begin{pmatrix} C \\ G \end{pmatrix}\begin{pmatrix} A \\ T \end{pmatrix}\right] = f\left[\begin{pmatrix} A \\ T \end{pmatrix}\right]f\left[\begin{pmatrix} C \\ G \end{pmatrix}\right]f\left[\begin{pmatrix} A \\ T \end{pmatrix}\right]f\left[\begin{pmatrix} G \\ C \end{pmatrix}\right] = \begin{pmatrix} T \\ A \end{pmatrix}\begin{pmatrix} G \\ C \end{pmatrix}\begin{pmatrix} T \\ A \end{pmatrix}\begin{pmatrix} C \\ G \end{pmatrix}$$

Assuming no mismatches, etc. then x and $f(x)$ shall be understood to represent the same molecule of duplex-stranded DNA.

Thus given $x = \begin{array}{cccc} A & G & C & T \\ T & C & G & A \end{array}$, then $f(x) = \begin{array}{cccc} A & G & C & T \\ T & C & G & A \end{array}$, and $x \cdot f(x)$ is a palindrome:

$$x \cdot f(x) = \begin{array}{cccccccc} A & G & C & T & A & G & C & T \\ T & C & G & A & T & C & G & A \end{array}$$

This works for odd lengths too.

Thus given $x = \begin{array}{cccc} A & G & C & T \\ T & C & G & A \end{array}$, then $x \cdot \begin{array}{c} G \\ C \end{array} \cdot f(x)$ is a palindrome:

$$x \cdot \begin{array}{c} G \\ C \end{array} \cdot f(x) = \begin{array}{ccccccccc} A & G & C & T & G & A & G & C & T \\ T & C & G & A & C & T & C & G & A \end{array}$$

Note: If we are given string $u c x f(c) v$, then $f(u c x f(c) v) = f(v) f^2(c) f(x) f(c) f(u) = f(v) c f(x) f(c) f(u)$, and $u c x f(c) v$ and $f(v) c f(x) f(c) f(u)$ represent the same molecule.

By the way, if w is a palindrome,
then $w = z \cdot f(z)$ for some z, thus $f(w) = f[z \cdot f(z)] = f^2(z) \cdot f(z) = z \cdot f(z) = w$
or $w = f(w)$.

Restriction Enzymes and Splicing Languages [b]

Let us list some cleavage sites of a few restriction enzymes. These enzymes break double-stranded DNA helices, and a few will be useful in understanding the relationships required for splicing languages (see Chapter 1).

Restriction Endonuclease	Cleavage Site
EcoRI	G A A T T C C T T A A G
TaqI	T C G A A G C T
SciNI	G C G C C G C G
HhaI	G C G C C G C G

Let us be given the following:

$$c = \binom{G}{C} \quad , \quad f(c) = \binom{C}{G} \quad , \quad x = \binom{A}{T}\binom{A}{T}\binom{T}{A}\binom{T}{A}$$

---

[b] We exclude restriction enzymes with required recognition sites such as HgaI (see Chapter 1).

Then given the two strings u c x f(c) v and p c x f(c) q, if we use EcoRI, as well as other requirements being satisfied (such as an energy source like ATP, and ligases), then we will obtain the following products in addition to the original strings:

u c x f(c) q   and   p c x f(c) v

Thus we have a way to generate new strings using restriction enzymes with ligases, and thus a language to describe such strings. Given such a scheme, we must take care that given overhangs (as well as blunt-end cleavage), only certain recombinations are permitted. Thus TaqI and SciNI recombinants are possible, but TaqI/HhaI, and SciNI/HhaI recombinations are not permitted.

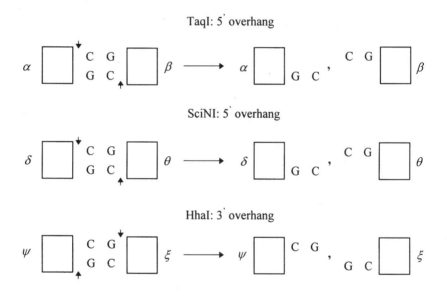

TaqI and SciNI strings may recombine, but TaqI/HhaI and SciNI/HhaI strings cannot recombine because they have different overhang types. The yield of valid hybrid strings are as follows (plus the original strings).

Viewed as blunt-ended strings, and $x = \begin{smallmatrix} C & G \\ G & C \end{smallmatrix}$, and $f(x) = x = \begin{smallmatrix} C & G \\ G & C \end{smallmatrix}$, then the hybrids formed are as follows: $\alpha\,xf(x)\beta$, $\delta\,xf(x)\theta \rightarrow \alpha\,xf(x)\theta$, $\delta\,xf(x)\beta$.

Notation: $\mathcal{N}$ will signify the set of restriction enzymes.

Linear Splicing System or an H System

Definition: A splicing system $S =_{df} (D, I, B, C)$ as follows,

D   finite alphabet

I   finite set of initial strings from D^*

B, C   pattern triples $< c, x, d >$, where c, x, d $\in D^*$, for all enzymes $\in$ N
    such patterns are called "sites"
    x is called a "crossing"
    B are left patterns   =   { 5$'$ overhangs } $\cup$ { blunt-end }
    C are right patterns   =   { 3$'$ overhangs }

A problem concerning representation of staggered-ended action exists. Given the
following, this problem will be elucidated.

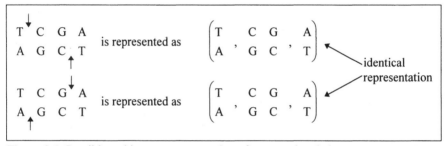

Figure 8.1 Possible ambiguous representation of staggered-end cleavage

As both 5$'$ and 3$'$ overhangs are not distinguished by this notation, instead two sets are
created, B and C. B holds the 5$'$ overhangs and blunt ends, while C holds the 3$'$
overhangs, and the distinctions are now preserved.

Example: For EcoRI $\in \mathcal{N}$,

$$G\!\downarrow\! A \; A \; T \; T \; C$$
$$C \; T \; T \; A \; A\!\uparrow\! G$$

5$'$ overhangs, thus $\in$ B

$$\begin{matrix} G\!\downarrow\! A \; A \; T \; T \; C \\ C \; T \; T \; A \; A\!\uparrow\! G \end{matrix} \quad \begin{matrix} \in \text{ B is equivalent} \\ \text{to the triple:} \end{matrix} \quad \left(\begin{bmatrix} G \\ C \end{bmatrix}, \begin{bmatrix} A \\ T \end{bmatrix} \begin{bmatrix} A \\ T \end{bmatrix} \begin{bmatrix} T \\ A \end{bmatrix} \begin{bmatrix} T \\ A \end{bmatrix}, \begin{bmatrix} C \\ G \end{bmatrix} \right) \in \text{B}$$

Example: For HhaI ∈ 𝓝,

$$\overset{\downarrow}{G}\ C\ G\ C$$
$$C\ G\ \underset{\uparrow}{C}\ G$$

3' overhang ,   thus ∈ C

$$\overset{\downarrow}{G}\ C\ G\ C$$
$$C\ G\ \underset{\uparrow}{C}\ G$$

∈ C equivalent
to the triple:

$$\left(\begin{bmatrix} G \\ C \end{bmatrix}, \begin{bmatrix} C \\ G \end{bmatrix} \begin{bmatrix} G \\ C \end{bmatrix}, \begin{bmatrix} C \\ G \end{bmatrix} \right) \in C$$

Example: For AluI ∈ 𝓝,

$$A\ \overset{\downarrow}{G}\ C\ T$$
$$T\ \underset{\uparrow}{C}\ G\ A$$

Blunt ends,   thus ∈ B

$$A\ \overset{\downarrow}{G}\ C\ T$$
$$T\ \underset{\uparrow}{C}\ G\ A$$

∈ B equivalent
to the triple:

$$\left(\begin{bmatrix} A \\ T \end{bmatrix} \begin{bmatrix} G \\ C \end{bmatrix}, 1, \begin{bmatrix} C \\ G \end{bmatrix} \begin{bmatrix} T \\ A \end{bmatrix} \right) \in B$$

Example: For HgiAI ∈ 𝓝,

$$G\ A\ G\ C\ \overset{\downarrow}{A}\ C$$
$$\underset{\uparrow}{C}\ T\ C\ G\ T\ G$$

3' overhangs ,   thus ∈ C

$$G\ A\ G\ C\ T\ \overset{\downarrow}{C}$$
$$\underset{\uparrow}{C}\ T\ C\ G\ A\ G$$

3' overhangs ,   thus ∈ C

$$G\ T\ G\ C\ \overset{\downarrow}{A}\ C$$
$$\underset{\uparrow}{C}\ A\ C\ G\ T\ G$$

3' overhangs ,   thus ∈ C

$$G\ T\ G\ C\ T\ \overset{\downarrow}{C}$$
$$\underset{\uparrow}{C}\ A\ C\ G\ A\ G$$

3' overhangs ,   thus ∈ C

the last example also has equivalent representation as triples, not included here.

Definition: $L = L(S) = \{ucxfq,\ \ pexd\ v\}$ then $\{ucxdv,\ \ pexfq\ \} \in L(S)$

and $<c, x, d>$, and $<e, x, f>$ are patterns of the same hand,

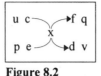

Figure 8.2

and $L = L(S)$ is the splicing language generated by the splicing system S.

Definition: Given a splicing system $S = (\ D, I, B, C\ )$, then we also have

$$S = (\ D, I \cup I^{'}\ , B, C\ ) \text{ where: } I^{'} = \{\omega \in D^{*}|\ \ f(\omega) \in I\}$$

A full example:

D	I	B	C
$S = (\{a, b, c, d, p, q, u, v, w, x\},$	$\{uaxbvaxbw,\ \ pcxdq\},$	$\{\langle a, x, b\rangle,\ \ \langle c, x, d\rangle\},$	$\varnothing\ )$

Figure 8.3

then $L(S) = (uax + pcx)(bvax)^{*}(dq + bw)$

(i.e., $\omega \in D^{*}$ then $f(\omega) \in D^{*}$ also, as ω and $f(\omega)$ are to represent the same molecule, and I is to represent initial strings, so must I). We will examine this example again very soon. Similarly, if $<c, x, d>$ is a pattern, then so must $f(<c, x, d>) = <f(d), f(x), f(c)>$.

The idea here is that we wish to generate all valid molecules that can be derived by splicing due to the valid application of restriction enzymes as represented by cleavages.

A few more examples to illustrate splicing systems:
Example: $S = (\ \{\ c, x\ \}, \{\ c\ x\ c\ x\ c\ \}, \{\ <c, x, c>\ \}, \phi\ )$, λ represents the null string

$c\ x\ c\ x\ c$ may be viewed as $\lambda \cdot (\ cxc\ ) \cdot x\ c$ and also as $c\ x \cdot (\ cxc\ ) \cdot \lambda$

since $p \cdot (\ cxd\ ) \cdot q$ and $u \cdot (\ cxd\ ) \cdot v$ implies $p \cdot (\ cxd\ ) \cdot v$ and $u \cdot (\ cxd\ ) \cdot q$

			result
then	step 1: $c\ x\ c\ x\ c$		$c\ x\ c\ x\ c$
	step 2: $\lambda \cdot (\ cxc\ ) \cdot \lambda$	$=$ $c\ x\ c$	$c\ x\ c,\ c\ x\ c\ x\ c,\ c\ x\ c\ x\ c\ x\ c$
	$\lambda \cdot (\ cxc\ ) \cdot xc$	$=$ $c\ x\ c\ x\ c$	
	$c\ x \cdot (\ cxc\ ) \cdot \lambda$	$=$ $c\ x\ c\ x\ c$	
	$c\ x \cdot (\ cxc\ ) \cdot x\ c$	$=$ $c\ x\ c\ x\ c\ x\ c$	
	step 3: using all strings up to and		$c\ x\ c\ x\ c\ x\ c\ x\ c,$
	including those in step 2,		$c\ x\ c\ x\ c\ x\ c\ x\ c\ x\ c$
	we get two more strings		

$L(S) = (\ c\ x\ )^{+} c = c\ (\ x\ c\ )^{+}$

Example: Let $d = \begin{pmatrix} C & G \\ G & C \end{pmatrix}$ and let $x = \begin{pmatrix} G & A & A & T & T & C \\ C & T & T & A & A & G \end{pmatrix}$ (Note: $f(x) = x$)

and consider d^{150} x d^{750} x d^{250}.

Let $a = d^{150}$, $b = d^{750}$, $c = d^{250}$,

and note that d is a palindrome, as are a, b, and c.

Let $I = \{ a x b x c \}$ be the initial set.

Note that $f(a x b x c) = f(c) f(x) f(b) f(x) f (a) = c x b x a$.

Let us concern ourselves with

$$S = (\{ a, b, c, x \}, \{ a x b x c, c x b x a \}, \{ x \}, \phi).$$

Then starting with	a x b x c	c x b x a
(and collecting terms)	a x a	c x a
	a x b x a	c x b x a
	a x b x b x a	c x b x b x a
	. . .	. . .
	a x c	c x c
	a x b x c	c x b x c
	a x b x b x c	c x b x b x c
	. . .	. . .

this continues (a fancy way to say use induction) and we obtain the following result:

$L(S) = (a + c)(x b)^* x (a + c) = (a + c) x (b x)^* (a + c)$

A description of this DNA language follows.

$$L(S) = \left(\left(\left(\begin{matrix} C & G \\ G & C \end{matrix}\right)^{150} + \left(\begin{matrix} C & G \\ G & C \end{matrix}\right)^{250}\right)\left(\begin{matrix} G & A & A & T & T & C \\ C & T & T & A & A & G \end{matrix}\right)\left(\begin{matrix} C & G \\ G & C \end{matrix}\right)^{750}\right)^*\left(\begin{matrix} G & A & A & T & T & C \\ C & T & T & A & A & G \end{matrix}\right)\left(\left(\begin{matrix} C & G \\ G & C \end{matrix}\right)^{150} + \left(\begin{matrix} C & G \\ G & C \end{matrix}\right)^{250}\right)$$

Note that the minimum length DNA double helix is $2 \times 150 + 6 + 2 \times 150 = 606$ bases long. In fact, four possible DNA strings can be formed, strings of the following lengths (in terms of bases):

$606 + 1506 \times n$

$806 + 1506 \times n$

$1006 + 1506 \times n$,   where: $n \geq 0$

Be aware that if artificial DNA is considered, then nucleotide base pairs such as iso C/iso G, κ/χ or κ/χ, α/π, β/δ, as well as chelated self complements such as hydroxypyridone, and phenylenediamine, as well as chelated base pairs such as dipicp/pyrimidine, and other nucleotide bases, mismatches, abasic pairs, etc. must be considered, as well. A review of Chapter 3 might prove useful. This remains true, even if endonucleases and ligases that apply to these alien nucleotide bases do not (currently) exist.

Thus the following are a possibility (at least for a chemist), although splicing systems currently have not been extended to deal with any nucleotide bases other than A, C, G, and T:

$$\left(\begin{matrix} C & G \\ G & C \end{matrix}\right)^{150}\left(\begin{matrix} \text{iso} & G & A & A & T & T & \text{iso} & C \\ \text{iso} & C & T & T & A & A & \text{iso} & G \end{matrix}\right)\left(\begin{matrix} C & G \\ G & C \end{matrix}\right)^{150}$$

and

$$\left(\begin{matrix} C & G \\ G & C \end{matrix}\right)^{150}\left(\begin{matrix} G & A & H & T & \emptyset & C \\ C & G & H & A & P & G \end{matrix}\right)\left(\begin{matrix} C & G \\ G & C \end{matrix}\right)^{150}$$

(mismatch, chelated self-complement, abasic pair)

Example: $S = (\{ c, x \}, \{ cxcxc \}, \{ <c, x, c> \}, \phi)$

initial   c x c x c:

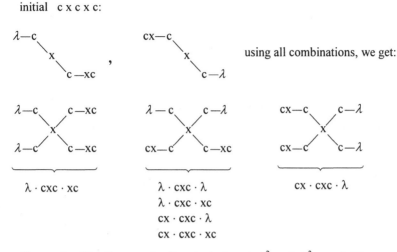

using all combinations, we get:

$\lambda \cdot cxc \cdot xc$

$\lambda \cdot cxc \cdot \lambda$
$\lambda \cdot cxc \cdot xc$
$cx \cdot cxc \cdot \lambda$
$cx \cdot cxc \cdot xc$

$cx \cdot cxc \cdot \lambda$

The result of the first set of splices is: $(cx)c + (cx)^2c + (cx)^3c \in L(S)$.

Using the new string generated, cx cx cx c, we create new splices:

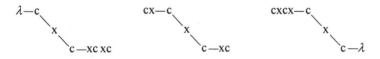

Using all the combinations of these splicings, we obtain the following result;
$(cx)c + (cx)^2c + (cx)^3c + (cx)^4c \in L(S)$. By induction, it is clear that

$L(S) = \{ (cx)^n c \mid n > 0 \} = \{ c (xc)^n \mid n > 0 \}$

If $<c, x, c>$ corresponds to

$$\begin{array}{l} A\ T{\downarrow}C\ G\ A\ T \\ T\ A\ G\ C_{\uparrow}T\ A \end{array}$$

where: $c = \begin{array}{l} A\ T \\ T\ A \end{array}$ and $x = \begin{array}{l} C\ G \\ G\ C \end{array}$

then $L(S) = \left\{ \begin{pmatrix} A & T & C & G \\ T & A & G & C \end{pmatrix}^n \begin{pmatrix} A & T \\ T & A \end{pmatrix} \;\middle|\; n > 0 \right\}$

Example: $S = \left(D, \ I, \ B, \ \varnothing \right)$

where:

$$D = \left\{ \binom{A}{T}, \ \binom{C}{G}, \ \binom{G}{C}, \ \binom{T}{A} \right\}$$

$$I = \left\{ \begin{bmatrix} A & G & A & T & C & T & & G & C & G & C & G & T & & G & G & A & T & C & C \\ T & C & T & A & G & A & & C & G & C & G & C & A & & C & C & T & A & G & G \\ \\ A & C & G & C & G & T & & G & G & A & T & C & T & & G & C & G & C & G & C \\ T & G & C & G & C & A & & C & C & T & A & G & A & & C & G & C & G & C & G \end{bmatrix} \right\}$$

$$B = \left\{ \begin{bmatrix} A & G & A & T & C & T & G & G & A & T & C & C \\ T & C & T & A & G & A & {}' & C & C & T & A & G & G \\ \\ A & C & G & C & G & T & G & C & G & C & G & C \\ T & G & C & G & C & A & {}' & C & G & C & G & C & G \end{bmatrix} \right\}$$

Then the resulting DNA language is as follows.

$$L(S) = \left\{ ab(cd)^{n} e \mid n \geq 0 \right\}$$

where: $a = \dfrac{\text{A C G C G T}}{\text{T G C G C A}}$ $b = \dfrac{\text{G G A T C T}}{\text{C C T A G A}}$

$c = \dfrac{\text{G C G C G T}}{\text{C G C G C A}}$ $d = \dfrac{\text{G G A T C T}}{\text{C C T A G A}}$

$e = \dfrac{\text{G C G C G C}}{\text{C G C G C G}}$

$S = (A, I, B, \phi)$

where: $A = \{ a, b, c \}$

$I = \{ baa, bb, aaac, cc, ac \}$

$B = \{ <b, a, \lambda>, <ba, a, \lambda>, <\lambda, b, \lambda>, <\lambda, aa, c>, <\lambda, aa, ac>, <\lambda, c, \lambda> \}$

Then so far, baa, bb, aaac, cc, ac, ba, b, aac, c $\in L(S)$ because if $< h, j, k > \in B$, then $\lambda \cdot h\,j\,k \cdot \lambda = hjk \in L(S)$.

- -

Since b, bb $\in L(S)$, and $<\lambda, b, \lambda>$, we have the following.

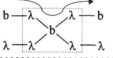

Following the arrow, we get b b b $\in L(S)$.

- -

Since bb, bbb $\in L(S)$, and $<\lambda, b, \lambda>$, we have the following.

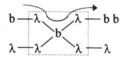

Following the arrow, we get b b b b $\in L(S)$.

Continuing this inductive process, we easily obtain the result that $b^+ \in L(S)$.

- -

Assuming that $b^+ \in L(S)$, and since ba, baa $\in L(S)$, we can continue this process to get $b^+aa \in L(S)$.

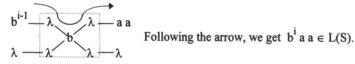

Following the arrow, we get $b^i a\,a \in L(S)$.

Continuing this inductive process, we easily obtain the result that $b^+a^* \in L(S)$.

- -

Since c, ac, aac, aaaac $\in L(S)$, and $<\lambda, aa, ac >, <\lambda, aa, c > \in B$, we have the following.

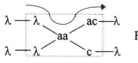

Following the arrow, we get a a a c $\in L(S)$.

(continued)

Let us assume a^i a a c $\in$ L(S) for some i, then

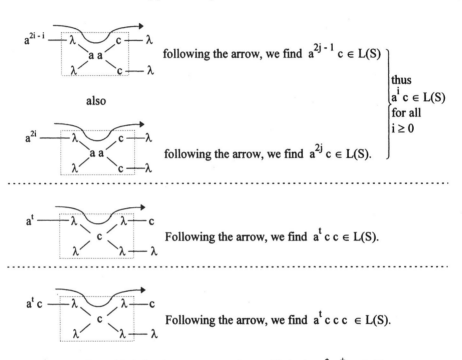

following the arrow, we find a^{2j-1} c $\in$ L(S)

also

following the arrow, we find a^{2j} c $\in$ L(S).

thus a^i c $\in$ L(S) for all i $\geq$ 0

Following the arrow, we find a^t c c $\in$ L(S).

Following the arrow, we find a^t c c c $\in$ L(S).

We can continue this inductive process, and we will obtain $a^* c^+ \in$ L(S).

Thus $L(S) = b^+ a^* + a^* c^+$.

Example: $S = \Big(A, \quad I, \quad B, \quad \varnothing \Big)$

         where:

$$A = \Big\{ a, \quad b, \quad c, \quad d, \quad p, \quad q, \quad u, \quad v, \quad w, \quad x \Big\}$$

$$I = \Big\{ uaxbvaxbw, \quad pcxdq \Big\}$$

$$B = \Big\{ \langle a, \quad x, \quad b \rangle, \quad \langle c, \quad x, \quad d \rangle \Big\}$$

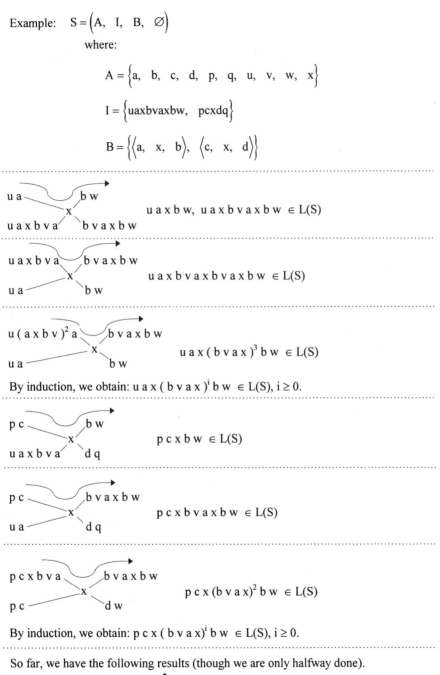

$u\,a\,x\,b\,w, \; u\,a\,x\,b\,v\,a\,x\,b\,w \; \in L(S)$

$u\,a\,x\,b\,v\,a\,x\,b\,v\,a\,x\,b\,w \; \in L(S)$

$u\,a\,x\,(b\,v\,a\,x\,)^3\,b\,w \; \in L(S)$

By induction, we obtain: $u\,a\,x\,(b\,v\,a\,x\,)^i\,b\,w \; \in L(S), \; i \ge 0.$

$p\,c\,x\,b\,w \; \in L(S)$

$p\,c\,x\,b\,v\,a\,x\,b\,w \; \in L(S)$

$p\,c\,x\,(b\,v\,a\,x)^2\,b\,w \; \in L(S)$

By induction, we obtain: $p\,c\,x\,(b\,v\,a\,x)^i\,b\,w \; \in L(S), \; i \ge 0.$

So far, we have the following results (though we are only halfway done).

$L(S) = (u\,a + p\,c)\,x\,(b\,v\,a\,x)^*\,b\,w$

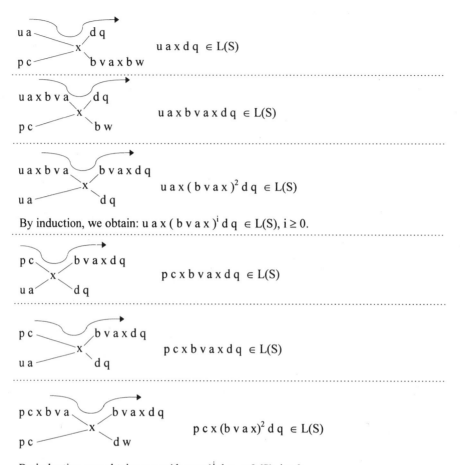

By induction, we obtain: u a x (b v a x)^i d q ∈ L(S), i ≥ 0.

By induction, we obtain: p c x (b v a x)^i d q ∈ L(S), i ≥ 0.

We get our answer by putting the above results together with the results from the previous page:

$$L(S) = (u a + p c) x (b v a x)^* (b w + d q)$$

Finite State Automata and Splicing Systems

A very good way to explain the finite state automaton approach to splicing systems. is by example. The reader is directed to this reading for more details ,[c] [75].

Definition: A splicing system $S = (A, I, B, C)$ is permanent if for each pair of strings a b x c d, e f x g h in A^* with sites $< b, x, c >, < f, x, h >$ of the same hand, then:
if y is a substring of a b x (respectively, x g h), that is, the crossing of a site in a b x c d (respectively, e f x g h), then substring y of a b x g h is the crossing of a site in a b x g h.

Definition: A splicing system $S = (A, I, B, C)$ is persistent if for each pair of strings a b x c d, e f x g h in A^* with sites $< b, x, c >, < f, x, h >$ of the same hand, then:
if y is a substring of a b x (respectively, x g h), that is, the crossing of a site in a b x c d (respectively, e f x g h), then substring y of a b x g h contains an occurrence of the crossing of a site in a b x g h.

If the splicing system S is a permanent splicing system, then an FSA may be constructed as exemplified in the following.

$$S = \left(A, \ I, \ B, \ \varnothing \right)$$

where: $A = \left\{ a, \ b, \ c, \ d, \ u, \ v, \ w, \ x, \ y \right\}$

$I = \left\{ udbcabv, \ abcvwabcd, \ vabxabcy \right\}$

$B = \left\{ \left\langle \lambda, \ ab, \ \lambda \right\rangle, \ \left\langle \lambda, \ bc, \ \lambda \right\rangle, \ \left\langle \lambda, \ d, \ \lambda \right\rangle \right\}$

It would be nice to point out instances of triples found in initial strings:

	string 1	string 2	string 3
	udbcabcv	abcvwabcd	vabxabcy
ab	udbcabcv	abcvwabcd	vabxabcy
bc	udbcabcv	abcvwabcd	vabxabcy
d	udbcabcv	abcvwabcd	--------

Figure 8.4

We can create a finite state automaton for each string, where each state is labeled "nk", where "n" refers to the string; thus state "23" is the fourth state for string 2.

---

[c] [62] the problem used is Example 2.2 on page 509, the method and application to this problem is on pages 515-516 (Theorem 4.1 with Lemmas 4.1.1 and 4.1.2, discussing Example 4.1).

Step 1

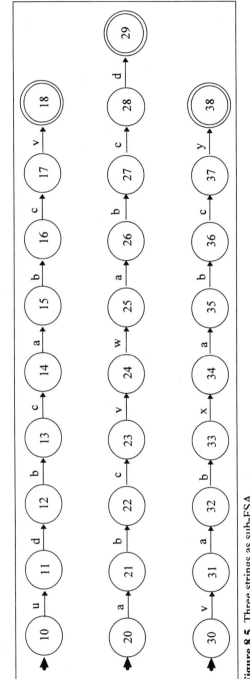

Figure 8.5 Three strings as sub-FSA

Step 2

"glue" or anneal common substring transitions together that appear as crossings. Specifically, for each crossing, glue states together as follows.

substring (crossing) ab: 16, 22, 27, 33, 36 becomes a composite state,
substring (crossing) bc: 14, 17, 23, 28, 37 becomes a composite state,
substrung (crossing) d: 12, 29   becomes a composite final state (29 is a final state)

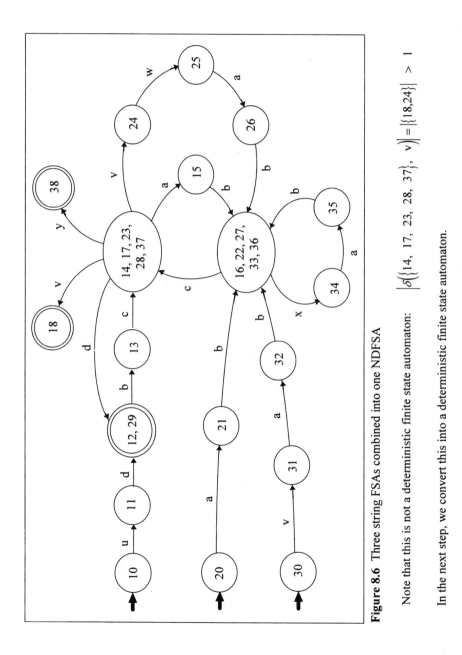

Figure 8.6 Three string FSAs combined into one NDFSA

Note that this is not a deterministic finite state automaton: $\quad \left\| \delta\left(\{14,\ 17,\ 23,\ 28,\ 37\},\ v\right) \right\| = \left\|\{18,24\}\right\| > 1$

In the next step, we convert this into a deterministic finite state automaton.

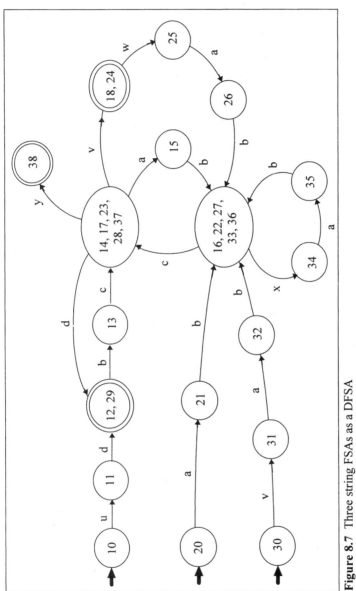

Figure 8.7 Three string FSAs as a DFSA

Note that this is now a deterministic finite state automaton.

In the next step, we obtain the regular expression that corresponds to this into a deterministic finite state automaton.

$$ud + \left[(v + \lambda)ab(xab)^* c + udbc\right]\left[(\lambda + vw)ab(xab)^* c + dbc\right]^* (d+v+y)$$

Insertion Sequences, Transposons, and Jumping Genes

From the point of view of notation, an insertion sequence is written as "IS" followed by a number used to identify it. Similarly, a transposon is written as "Tn" followed by a number used to identify it. An insertion sequence is a segment of genetic material of the order of two kilobases [26, p. 7], and can be transferred from a virus (phage) to a plasmid in a bacterium, and then from such a bacterium to yet a higher organism. Insertion sequences are relatively short sequences of base pairs, but do not contain genes. Insertion sequences may attach themselves or insert themselves into genes, and by interrupting the sequences of base pairs in genes, they modify the gene expression, usually by turning off the gene's expression. A transposon is much larger than an insertion sequence and may contain genes. Thus when a transposon inserts itself within the base pairs of any gene, it turns off that gene's expression, and replace that gene's expression with the expression of the genes that are carried within the transposon. Just as with an insertion sequence, transposons may move from a bacterial virus, to a plasmid in a bacterium, or into a higher organism. Transposons are sometimes referred to as "jumping genes" as they have the ability to jump or change position from one section of DNA to another position on the same DNA segment (changing genetic expression as they do so). An example of this "transposon" action is the ability to transfer antibiotic resistance to bacteria. Thus the important thing is that target t is replaced by t (X) t below (t being a direct repetition).

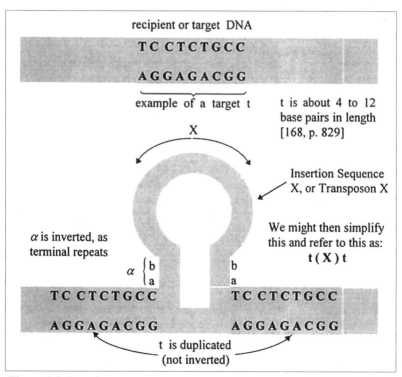

Figure 8.8 Insertion sequence or transposon

Nomenclature [26, pp. 15-22][d]

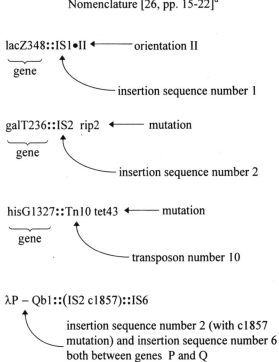

λP – Qb1::(IS2 c1857)::IS6

        insertion sequence number 2 (with c1857
        mutation) and insertion sequence number 6
        both between genes P and Q

---

[d] Elucidation of these comments, courtesy E. Lederberg.

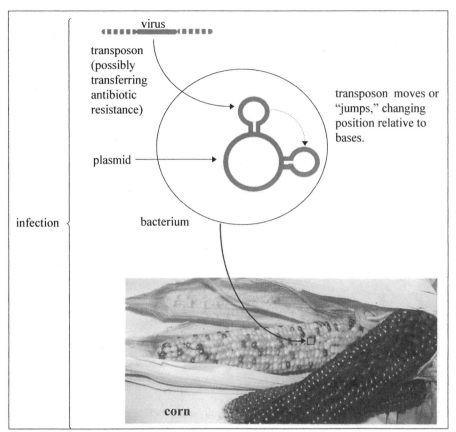

Figure 8.9

Note: Let us assume the following takes place on a plasmid. A transposon is incorporated at site number 1, by disrupting the gene number 1 located at that site, gene number 1 is "turned off," and transposon gene "A" might be turned on. If or when the transposon "jumps" to a new site number 2, the gene number 1 may now be turned on again, but the transposon may be incorporated at a new site number 2 (turning off this gene), but possibly activating transposon gene "B" instead of gene "A": thus a sequence of genetic expression of genes turning on and off.

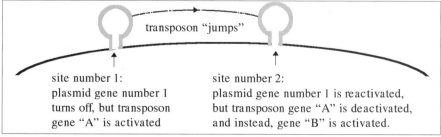

Figure 8.10

Head [64] characterizes splicing systems and concludes that such systems generate strictly locally testable or regular expressions. Gatterdam [62] points out that if a Splicing system S = (D, I, B, C) is such that there are no patterns < a, x, b > ∈ B and < c, x, d > ∈ C for the same crossing x, then S is crossing disjoint. Using crossing disjoint Splicing systems, Gatterdam points out that L(S) may be proven to be regular by construction.

Upon allowing for insertion sequences or transposons to insert base sequences or to excise base sequences [64, p. 754], [65], Head attempts to characterize splicing systems with these new operations.

   Let i be an IS or Tn
   Let t be a target (for insertion of an IS or Tn)

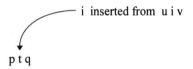

i inserted from u i v

p t q

Specifically, Head states that p t q and u i v support the result p t i t q.

To support such a view, Head states that we require a production rule of the form

   t ⇒ t (t) t

Head states that under this interpretation, the languages that result are context-free and are no longer regular.

Circular DNA and Splicing Systems

Circular splicing systems generalize linear splicing systems. Being limited to linear splicing systems is acceptable computationally, but in the world of biology, biochemistry, and genetics, etc., such limitations are not found. The previous discussion about linear splicing systems has been studied by others, including papers that the reader can investigate [43], [44]. We now examine circular splicing systems.

Notation: $c = {}^\wedge a_1 \ldots a_n$ will signify a circular string of symbols, where $a_i \in D^*$,

and $a_1 \ldots a_n$ represents one of the equivalent linearized forms of the circular string c .

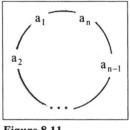

Figure 8.11

The equivalence relation is $\sim$ and all cyclic permutations of c are equivalent.

Thus if c = abbac with $D = \{ a, b, c \}$, then the following strings of c are equivalent:
abbac $\sim$ cabba $\sim$ acabb $\sim$ bacab $\sim$ bbaca

Two sets of inverse string operations are required to deal with circular DNA, referred to as SA1, SA2, SA3, and SA4.

For definitions, see [167] and [65].

A circular splicing system $S = (D, T, P, I, J)$ where:

D	is a finite vocabulary
T	pattern triples $< x, y, z >$, where x, y, z $\in D^*$
I	is a finite set of initial linear strings
J	is a finite set of initial circular strings
P	is the "pairing" relationship, defined as follows:

$$< p, x, q > P < u, y, v > \quad \text{then} \quad x = y$$

for linear strings,

given h p x q k and w u x v z and $< p, x, q > P < u, x, v >$, then

h p x v z   and   w u x q k

for circular strings,

given $\hat{}$ h p x q and $\hat{}$ w u x v   and $< p, x, q > P < u, x, v >$, then

$\hat{}$ h p x v w u x q

repeats
(review transposons)

Alternate definitions are used by different investigators, an example being found in [137]. Pixton generalizes, as he is very concerned with reflexivity, especially with regard to self-splicing. His generalization of splicing systems is more generalized as he feels that the definitions used here are "... more adapted to biochemical applications...," a limitation which we welcome.

$$L(S) = \left\{ w \mid w \in \hat{}\,D^*, \ J \underset{SA1, \, SA2}{\overset{*}{\Rightarrow}} w, \ I = \varnothing \right\}$$

$$\cup \left\{ w \mid w \in D^* \cup \hat{}\,D, \ J \underset{SA3}{\overset{*}{\Rightarrow}} w \right\}$$

$$\cup \left\{ w \mid w \in D^* \cup \hat{}\,D, \ I \underset{SA4}{\overset{*}{\Rightarrow}} w, \ J = \varnothing \right\}$$

The definitions of SA1, SA2, SA3, and SA4 now follow.

Given {^hpxq, ^wuxv}, then {^hpxvwuxq}

or given two circular strings sharing crossing x, we may construct a circular string composed of both of the original circular strings (a closure operation).

This operation is named SA1.

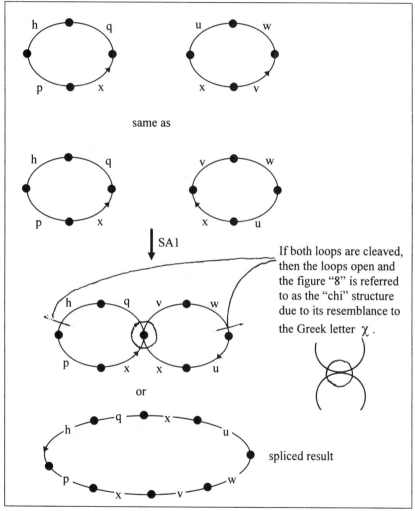

Figure 8.12

The example appears as Figure 32.5 [95, p. 631].

Given {^hpxqwuxv}, then {^hpxv, ^wuxq}

or given a circular string with two instances of crossing x, we obtain two circular strings each having an instance of crossing x.

This operation is named SA2 and is the inverse of opertion SA1.

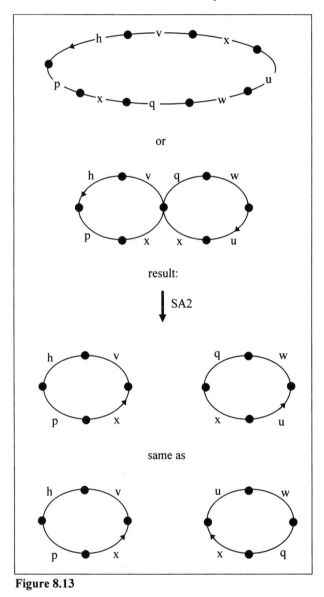

Figure 8.13

The example appears as Figure 32.6 [95, p. 631].

Given {hpxqk, ^wuxv}, then {hpxvwuxqk}

or given a linear string with a circular string, both sharing crossing x, we may construct a new linear string composed of both the original linear string followed by the circular string with two copies of crossing x.

This operation is named SA3.

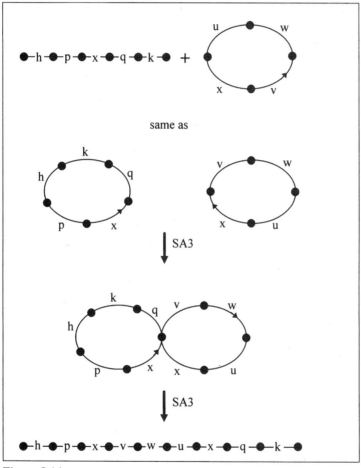

Figure 8.14

It is worthwhile examining this a little more closely.

Let us assume that the crossing x is as follows: $x = \begin{matrix} \text{G C T T T T T T A T A C T A A} \\ \text{C G A A A A A A T A T G A T T} \end{matrix}$

What does this "splice" of circular DNA really look like?
The example appears as Figure 32.19 [95, p. 643].

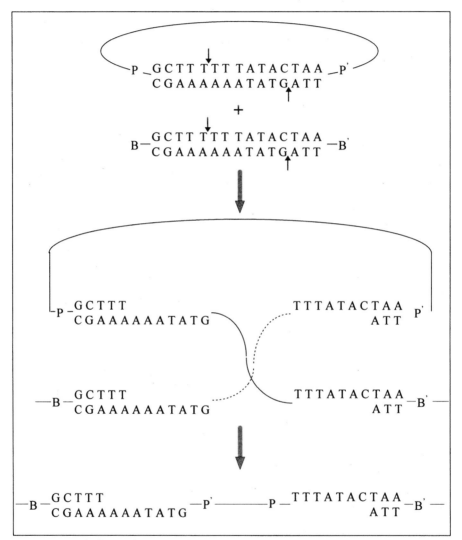

Figure 8.15 Circle opening into a linear string, via staggered ends

Given {hpxqkuxvz}, then {hpxvz, ^kuxq}

or given a linear string with two instances of crossing x, we obtain a new linear string with crossing x, and a circular string with crossing x.

This operation is named SA4 and is the inverse of operation SA3.

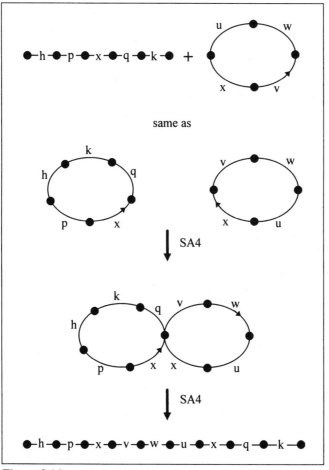

Figure 8.16

The example appears as Figure 32.18 [95, p. 642] and Figure 33.9 (Transposons with direct repeats) [95, p. 657].

Not all regular languages are splicing languages. To be explicit, $(aa)^*$ is not a splicing language.

Assume that $(aa)^*$ is a splicing language, then there must exist a pattern $< a^p, a^q, a^r > \in T$.
If $p + q + r$ is even:
   then $a^p \cdot a^q \cdot a^r \in (aa)^*$, thus $a \cdot a^p \cdot a^q \cdot a^r \cdot a \in (aa)^*$, but then $a^p \cdot a^q \cdot a^r \cdot a \notin (aa)^*$, but this is a contradiction if we have a splicing language.
If $p + q + r$ is odd:
   then $\lambda \cdot a^p \cdot a^q \cdot a^r \cdot a \in (aa)^*$, and also $a \cdot a^p \cdot a^q \cdot a^r \cdot \lambda \in (aa)^*$,
      but then $a^p \cdot a^q \cdot a^r \notin (aa)^*$, but this is a contradiction if we have a splicing language.
$P + q + r$ must be even or odd, and no matter which, we get a string that leads to a contradiction if we assume that $(aa)^*$ is a splicing language; thus the assumption that $(aa)^*$ is a splicing language is false [60, p. 65].

Using the fact that $(aa)^*$ is not a linear splicing language, we conclude that there exist circular splicing languages that cannot be generated by any linear splicing system using SA1.

Given that $L = (aa)^*$ is not a linear splicing language,
   then $L_1 = \{ (aa)^n \mid n \geq 0 \}$ can be generated using SA1 by the circular splicing system ($\{a\}$, $\{<\lambda, a, \lambda>\}$, $< \lambda, a, \lambda > P < \lambda, a, \lambda >$, ϕ, $\{ \lambda, aa \}$):

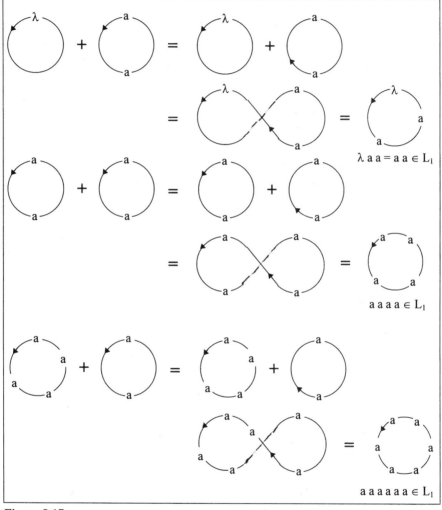

Figure 8.17

Note: The language $L = \{a^n ba^n \mid n > 0\}$ is a well-known context-free language, but if this language is viewed as composed of circular strings, then

$$^\wedge L = \{^\wedge a^n ba^n \mid n > 0\} = \{^\wedge a^{2n} b \mid n > 0\}$$ which is well known to be regular.

$$a^n b a^n \sim a^n a^n b = a^{2n} b$$

$$L = \left\{ ^\wedge (ca)^n (cb)^n \mid n > 0 \right\}$$ using SA1 is context-free, given the following.

Given $< ca, c, b >, < cb, c, a > \in T$, then $ca\,c\,a$, and $cb\,c\,b$ are valid strings, thus:

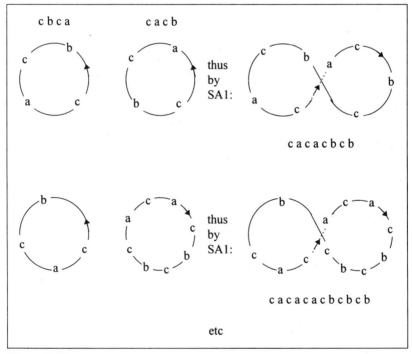

Figure 8.18

$$L = \left\{ {}^{\wedge} a^n b \mid n > 0 \right\}$$

L is a regular circular string language. Using operation SA1, L is not closed (is not a regular circular string langage).

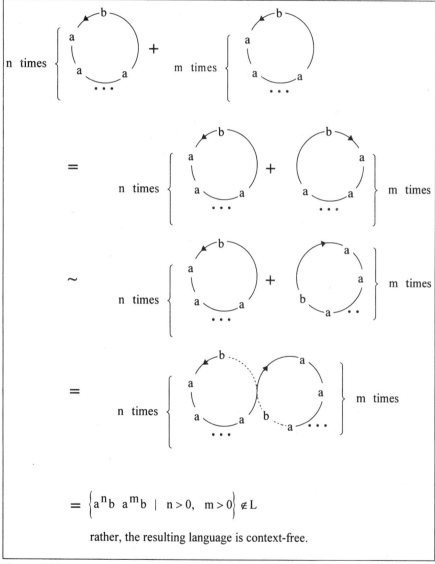

$$= \left\{ a^n b \; a^m b \mid n > 0, \; m > 0 \right\} \notin L$$

rather, the resulting language is context-free.

Figure 8.19

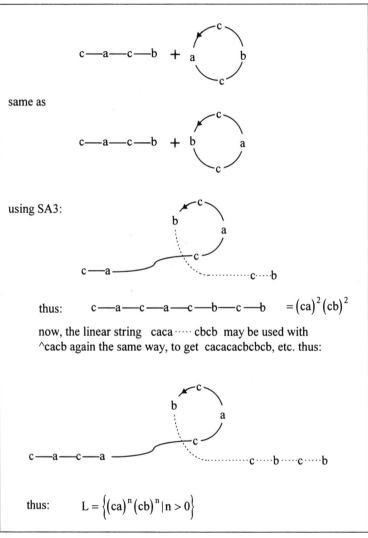

same as

using SA3:

thus: c —a —c —a —c —b —c —b $= (ca)^2 (cb)^2$

now, the linear string   caca ····· cbcb   may be used with
^cacb again the same way, to get cacacacbcbcb, etc. thus:

thus: $L = \left\{ (ca)^n (cb)^n \mid n > 0 \right\}$

Figure 8.20

There exist context-free string languages generated by circular string systems using SA3.

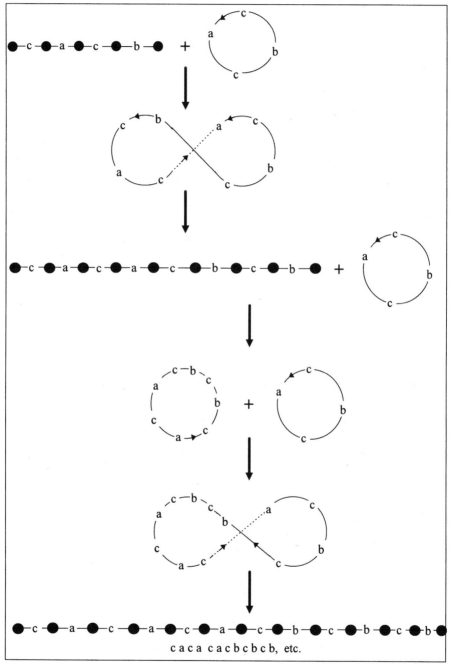

Figure 8.21

Hoogsteen Complements vs. Watson-Crick Complements

Triple and quadruple DNA helices are often associated with genetic diseases such as HPFH [47, p. 54]. Quadruple helices (G-quartets, for example) are components of telomeres, and are used in anti-carcinogenic pharmaceuticals [47].

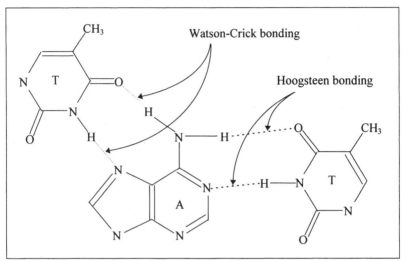

Figure 8.22 Triple-Stranded DNA

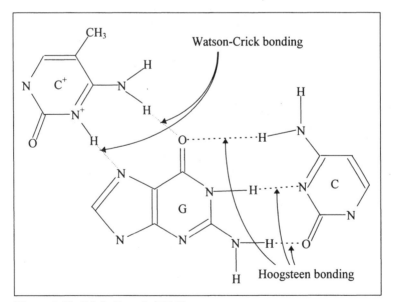

Figure 8.23 Triple-Stranded DNA

See [46].

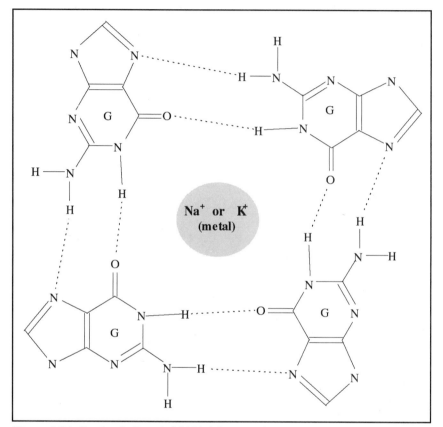

Figure 8.24 Quadruple Stranded DNA (G-quartet)

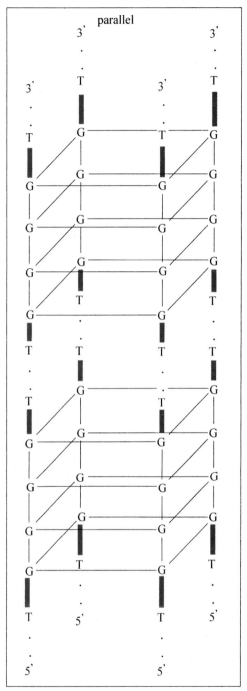

Figure 8.25 G Quartet, see [47, pp. 53, 54].

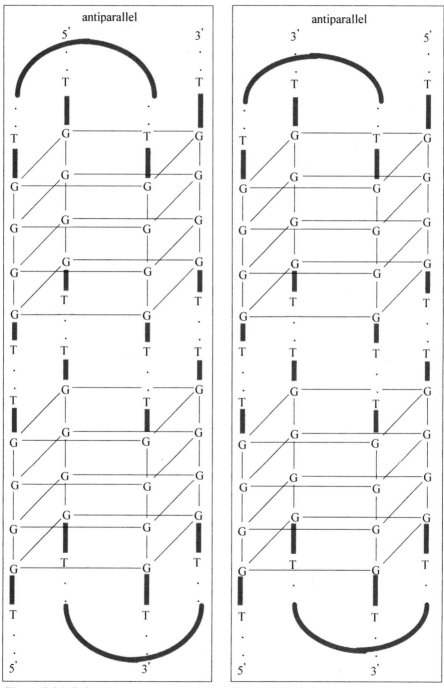

Figure 8.26 G Quartet

Figure 8.27 G Quartet

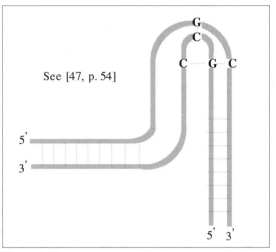

Figure 8.28

DNA triple helices such as this example will be drawn in schematic form as follows.

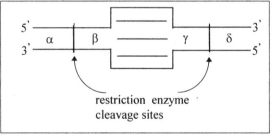

Figure 8.29

As restriction enzymes tend to cleave at the junction of double and triple helices, the following are possible:

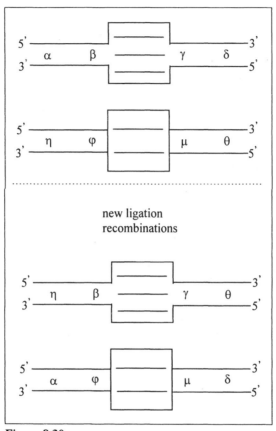

Figure 8.30

The computational studies in the literature have not discussed triple or quadruple helices (or at least, not to any large extent), but the above example indicates at least some of the possibilities when mixing double- and triple-stranded helices with restriction enzymes followed by ligation. Splicing systems (both linear as well as circular) have not taken into account such mixed double and triple DNA helices.

We must now take a moment to discuss "dominoes," but we will resume this discussion.

Dominoes [43], [44]

Dominoes are a way to formalize the description of splicing. The brief discussion that follows is a good introduction to this subject.

Definition: A Δ-domino α is a triple, as follows: α = < $l(\alpha)$, $m(\alpha)$, $r(\alpha)$ >. These components of α are referred to as the domino's left, middle, and right parts.

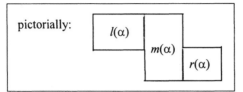

Figure 8.31

Note: If α = $m(\alpha)$, then α is a blunt domino.

Two dominoes may be matched (ligated) to form a larger Δ-domino. Matching is a binary relation, written as ⊗. A graphical representation will make this clear.

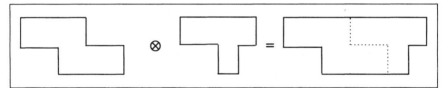

Figure 8.32

Note that if α, β ∈ Δ (we have two dominoes), then:

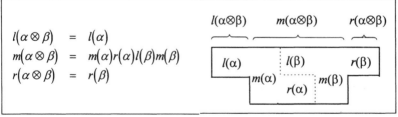

$$l(\alpha \otimes \beta) = l(\alpha)$$
$$m(\alpha \otimes \beta) = m(\alpha)r(\alpha)l(\beta)m(\beta)$$
$$r(\alpha \otimes \beta) = r(\beta)$$

Figure 8.33

If DNA is double-stranded (thus we are restricted to double-stranded DNA), then a couple of strings can represent the upper and lower strands. Thus given (x, y), x is a string at the upper strand and y is the string at the lower strand (of course, Watson-Crick complementarity must be respected–or to be explicit, mismatches are excluded). Thus:

$$\ldots x \ldots$$

$$\ldots y \ldots$$

Thus given the domino

$$\alpha = <(\text{AG}, \lambda)(\text{G T C}, \text{C A G})(\text{T}, \lambda)> = < l(\alpha), m(\alpha), r(\alpha) >$$

then each couple represents parts of both upper and lower strands, where:

$$l(\alpha) = (\text{AG}, \lambda) \quad = \quad \frac{\text{AG}}{\lambda}$$

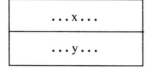

$$m(\alpha) = (\text{G T C}, \text{C A G}) = \frac{\text{G T C}}{\text{C A G}} \quad \text{and} \quad \alpha =$$

$$r(\alpha) = (\text{T}, \lambda) \quad = \quad \frac{\text{T}}{\lambda}$$

Another example is provided by

$$\alpha = <(\text{AG}, \lambda)(\text{G T C}, \text{C A G})(\text{T}, \lambda)> = < l(\alpha), m(\alpha), r(\alpha) >$$

$$\beta = <(\lambda, \text{A})(\text{A}, \text{T})> = < l(\beta), m(\beta), \lambda >$$

thus $\alpha \otimes \beta = < l(\alpha \otimes \beta), m(\alpha \otimes \beta), r(\alpha \otimes \beta) >$, where:

$$
\begin{aligned}
l(\alpha \otimes \beta) &= l(\alpha) &&= (\text{AG}, \lambda) \\
m(\alpha \otimes \beta) &= m(\alpha)\, r(\alpha)\, l(\beta)\, m(\beta) &&= (\text{G T C}, \text{C A G})(\text{T}, \lambda)(\lambda, \text{A})(\text{A}, \text{T}) \\
r(\alpha \otimes \beta) &= r(\beta) &&= (\lambda, \lambda)
\end{aligned}
$$

$$\alpha \otimes \beta =$$

In the literature, dominoes are discussed from the viewpoint of semigroups. Languages can be viewed as taking place over (free) semigroups, string concatenation being the binary operation, with λ being the monoid identity.

If the semigroup is S, then $\Delta \subset S \times S$, and:

$$l(\alpha),\ r(\alpha) \in \{ \text{domain } \Delta \times \lambda \} \ \cup \ \{ \lambda \times \text{range } \Delta \} \ \cup \ \{ \lambda \times \lambda \}, \text{ and } m(\alpha) \in \Delta$$

We shall not explore the semigroup viewpoint any further; the interested reader is encouraged to do so.

However, before proceeding further, talk about upper strand and lower strand might be viewed as too informal, so projection operators are provided to take care of such informality. Thus

$\tau_1 : \alpha \rightarrow$ upper strand

$\tau_2 : \alpha \rightarrow$ lower strand

and if given $\alpha = (x, y)$, then more formally, we have

$$\alpha = \left(\tau_1(\alpha), \ \tau_2(\alpha) \right)$$

Oriented Dominoes

If σ represents an oriented domino, then to be an oriented domino,

$l(\sigma) = (l, \lambda)$ but not (λ, λ) implies $r(\sigma) = (\lambda, r)$ or (λ, λ)

or

$l(\sigma) = (\lambda, l)$ but not (λ, λ) implies $r(\sigma) = (r, \lambda)$ or (λ, λ)

Thus we obtain

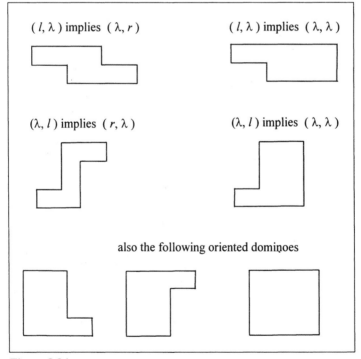

Figure 8.34

The product of two oriented dominoes α and β

where: σ is the oriented product of α and β is now written as $|_P$

Thus $\sigma = \alpha \ |_P\ \beta$ and $\quad m(\sigma) \ = r(\alpha)\, l(\beta)$

$$l(\sigma) \quad = (\lambda, x), \text{ where x is a suffix of } \sigma$$

$$r(\sigma) \quad = (y, \lambda), \text{ where y is a prefix of } \sigma$$

We are now ready to use some of the machinery developed:

$$\alpha_1 = (a^2, a^2)(ab^2, \lambda), \quad \alpha_2 = (\lambda, ab^2)(b, b), \quad \alpha = (a^3b^3, a^3b^3)$$

$$(a^2, a^2)(ab^2, \lambda) \ |_P\ (\lambda, ab^2)(b, b) = (a^3b^3, a^3b^3) \ = \alpha$$

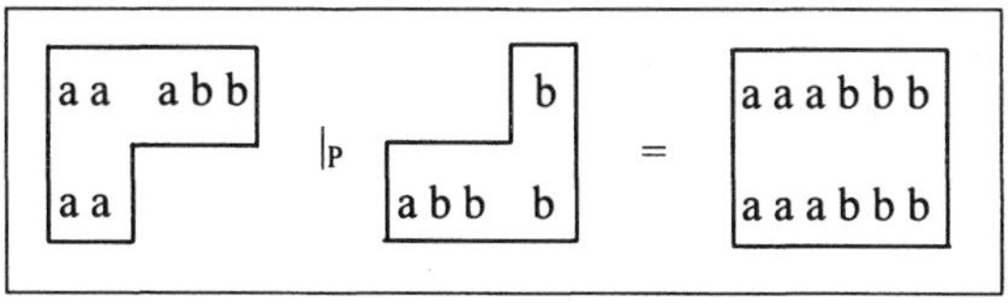

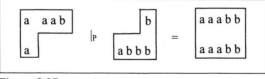

Figure 8.35

$$\alpha_3 = (a, a)(a^2b, \lambda), \quad \alpha_4 = (\lambda, a^2b)(b^2, b^2)$$

$$(a, a)(a^2b, \lambda) \ |_P\ (\lambda, a^2b)(b^2, b^2) = (a^3b^3, a^3b^3) \ = \alpha$$

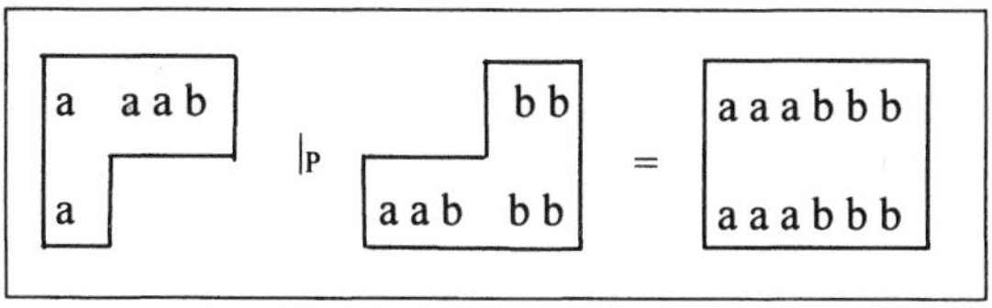

Figure 8.36

Note: $\alpha_3 \ |_P\ \alpha_2 \ = (a^3b^2, a^3b^2)$

Figure 8.37

Now we return to our discussion concerning triple- and higher-stranded helices of DNA. What we find is that as currently constituted, dominoes have been designed to be limited to double-stranded helices of DNA. This is not to imply, however, that dominoes cannot be redefined to support the more complicated structures found in triple-stranded DNA, however. For a quick investigation of some of the cases where dominoes would have to be changed to enable support of triple-stranded DNA, see Figures 38 and 40; Figure 39 represents an actual experimental case.

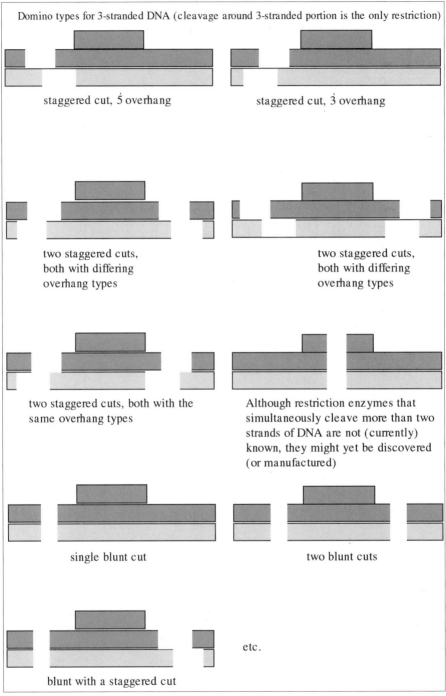

Domino types for 3-stranded DNA (cleavage around 3-stranded portion is the only restriction)

staggered cut, 5̇ overhang

staggered cut, 3̇ overhang

two staggered cuts,
both with differing
overhang types

two staggered cuts,
both with differing
overhang types

two staggered cuts, both with the
same overhang types

Although restriction enzymes that
simultaneously cleave more than two
strands of DNA are not (currently)
known, they might yet be discovered
(or manufactured)

single blunt cut

two blunt cuts

blunt with a staggered cut

etc.

Figure 8.38

```
                      E
                                    Me    MeMe    Me
                                    TTTTCTTTTTCCTTTTCTTTT————3'
5'———NTTTTCTTTTCCTTTTCTTTT    ACTAGT    AAAAGAAAAGGAAAGAAAAN————3'
3'———NAAAAGAAAAGGAAAGAAAA     TGATCA    TTTTCTTTTTCCTTTTCTTTTN————5'
3'———TTTTCTTTTTCCTTTTCTTTT
     Me  Me Me   Me                E

     N   any nucleotide base (or its complement)
     E   an electrophile
     Me  methyl group
```

Figure 8.39

Triple-stranded DNA (above) causes double-stranded DNA cleavage. This is an artificially designed restriction nuclease that is effective at 85% of the sites, and the cleavage points may be ligated [46, p. 5937].

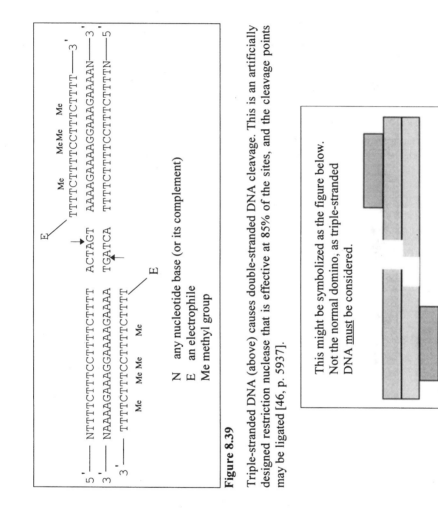

This might be symbolized as the figure below. Not the normal domino, as triple-stranded DNA <u>must</u> be considered.

Figure 8.40

Splicing Systems on Graphs

A more detailed analysis [53] of the mechanisms of restriction enzymes, modeled by graphs, appears to have the possibility of supporting more complicated types of splicing. This view is based upon a graph-theoretic approach to DNA and restriction enzyme action. Currently, restriction enzymes are described in the following way; an example is sufficient to explain a graph-theoretic approach

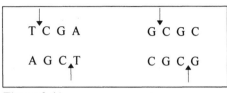

Figure 8.41

We understand by the above that this is an abbreviated way to indicate how molecular bonds are cleaved, not by examining the details of covalent bonds, but at a higher level. We understand the above to mean (with a little greater detail), the $5'$ and $3'$ bond structures as follows.

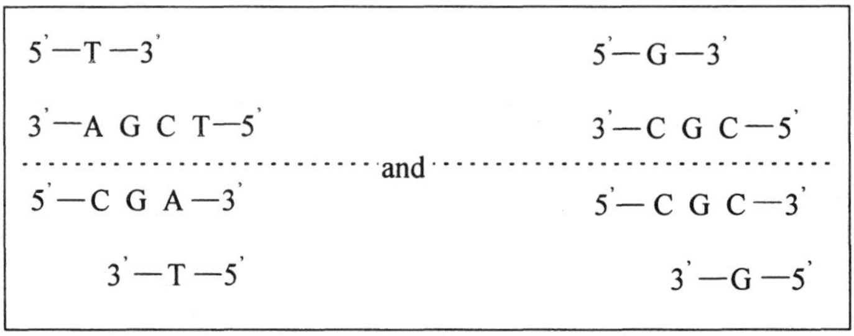

$$5'—T—3' \qquad\qquad\qquad 5'—G—3'$$

$$3'—A\ G\ C\ T—5' \qquad\qquad 3'—C\ G\ C—5'$$

.... and ....

$$5'—C\ G\ A—3' \qquad\qquad 5'—C\ G\ C—3'$$

$$3'—T—5' \qquad\qquad\qquad 3'—G—5'$$

Figure 8.42

Recall, at the end of the first chapter (review of biochemistry), it was noted that DNA/RNA hybrids exist, with cleavages such as the following.

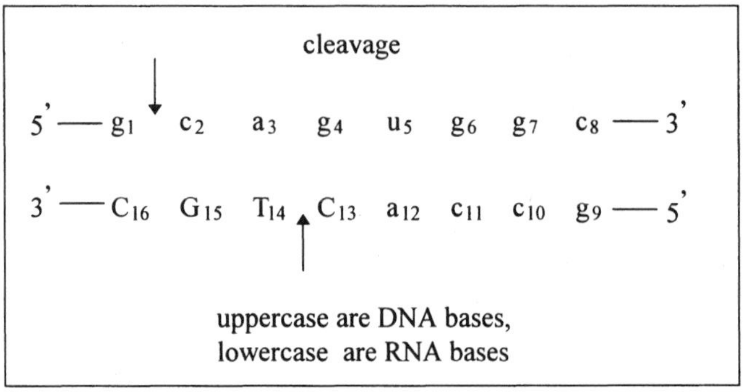

cleavage

$$5'\ —\ g_1 \quad c_2 \quad a_3 \quad g_4 \quad u_5 \quad g_6 \quad g_7 \quad c_8\ —\ 3'$$

$$3'\ —\ C_{16} \quad G_{15} \quad T_{14} \quad C_{13} \quad a_{12} \quad c_{11} \quad c_{10} \quad g_9\ —\ 5'$$

uppercase are DNA bases,
lowercase are RNA bases

Figure 8.43

Thus domino and splicing systems may have to be expanded to include such cleavages as well (assuming that ligases can act upon these cleavages).

Now, for the graph-theoretic approach.

Graph-Theoretic Approach to Splicing Systems

By using more information to describe the molecular structure, expressed in graph-theoretic terms, a closer modeling can be attained in splicing systems. This graph-theoretic approach has much promise, especially as it appears capable of describing more complex kinds of splicing such as appears with triple-helix DNA. An example of this graph-theoretic approach follows. However, just how versatile is this method? More work must be done to determine the answer to this question!

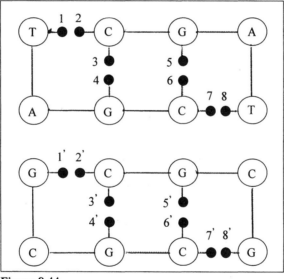

Figure 8.44

Thus by providing a list of connections from one node to another (which models the bonding), we can specify (in graph-theoretic form) how the splice will take place; thus,

$$\{ (1, 2'), (3', 4), (5', 6), (7, 8') \} \cup \{ (1', 2), (3, 4'), (5, 6'), (7', 8) \}$$

Linear Splicing on Graphs

Example:

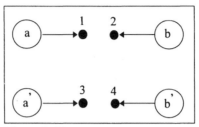

Figure 8.45

with the splicing scheme of $\{(1,4), (3,2)\}$ yields:

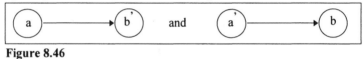

Figure 8.46

thus $u\,a\,b\,v$ and $u'\,a'\,b'\,v'$ yields $u\,a\,b'\,v'$ and $u'\,a'\,b\,v$

Circular Splicing on Graphs

Example:

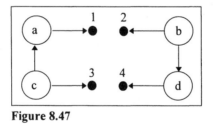

Figure 8.47

with the splicing scheme of $\{(1,4), (3,2)\}$ yields:

Figure 8.48

Another example:

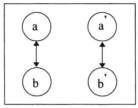

Figure 8.49

with the splicing scheme of { (a, b'), (a', b) } yields:

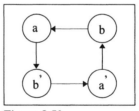

Figure 8.50

For the interested reader, there is a series of technical articles that may be consulted by such authors as G. Paūn and G. Rozenberg as well as a book *DNA Computing: New Computing Paradigms* by G. Paūn , G. Rozenberg, and A. Salomaa which is especially concerned with the interrelationships between splicing systems, sticker systems, H systems, and Watson-Crick automata, emphasizing their computational or mathematical abstractions.

Chapter 9: tRNA Structure

Plague

"The Magistracy's concern over possible noncompliance or downright hostility from ecclesiastics with respect to health ordinances and officials was not unfounded. Particularly in times of epidemics, grounds for conflict between Public Health officials and men of the Church could appear at any moment, and in the most unexpected quarters" [223, p. 3].

"The most frequent sources of conflict between clergymen and health officers were: the quarantine measures to which the clergy were not always willing to submit; the requisition of monasteries or other religious premises which the officers needed to transform into temporary pesthouses or places for convalescents; and last, but not least, the sermons and processions. During epidemics the health officers did not look favourably upon gatherings of people" [223, p. 6].

"As far as sermons and processions were concerned, there is no doubt that most of the health officers shared the religious convictions of their time, and felt that the sermons and processions might appease God, and thus perhaps help to bring the epidemic to an end. But while the appeasement of God's wrath through sermons and processions appeared to the health officers as a possibility, the exacerbation of contagion through large assemblies of people was for them a certainty" [223, p. 7].

tRNA Structure and Knot Theory

tRNA (mRNA, DNA, etc.) may have a complicated structure, analyzed by primary, secondary, tertiary, etc. structure. Folding is not an accidental artifact, but serves a very important function: areas that should rarely be decoded may be shielded by folding, while areas that should frequently be decoded can receive maximal exposure. The secondary structure is complicated, however. Thus the question is, how can this complicated structure be disentangled? The methodology of analyzing such complicated, folded, three-dimensional structures includes the possibility of employing knot theory. Knot theory may thus be applied in the theory of computation apart from the already mentioned area involving topoisomerases (Chapter 1). Although we shall not dwell long upon this study, a few papers have proven to be very worthwhile studying and are intrinsically interesting. We shall first study the ideas involved in a Nussinov plot [318] and then see how these ideas are related to knot theory.

We assume that the RNA secondary structure is topologically equivalent to a plane (has no knots or pseudoknots: thus there are no intersecting edges and no bases are paired more than once). Thus, if RNA is viewed as planar, it might have the cloverleaf appearance below. The Nussinov plot allows us to determine this overall secondary structure. Easy for us to see, once we have it! The question is: How do we obtain the structure in Figure 9.1, below?

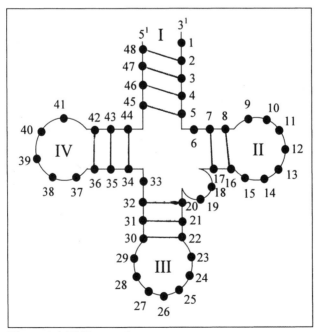

Figure 9.1 RNA secondary structure

Nussinov Plots

The Nussinov plot (Figure 9.2) is determined by placing the bases on a circle, and connecting the paired bases with chords. As is easily seen, there are four areas of loops (also called flowers) that correspond to the four subregions seen in Figure 9.1.

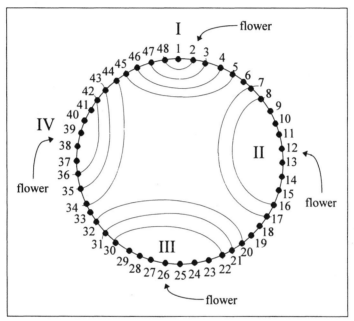

Figure 9.2 Example of a Nussinov plot for RNA in Figure 9.1

There is a restriction, however. Chord bases B_i and B_j may not be adjacent (this condition may be relaxed, however) [318, p. 74].

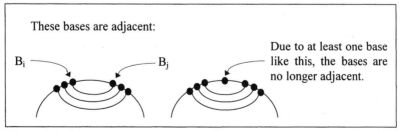

Figure 9.3 Adjacent bases

Given the following Nussinov plot, can we predict the corresponding tRNA secondary structure?

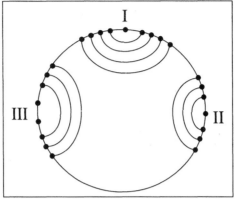

Figure 9.4

The predicted secondary structure:

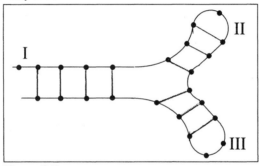

Figure 9.5

More complicated structures are possible, however. Let us examine the Nussinov plot for the following structure.

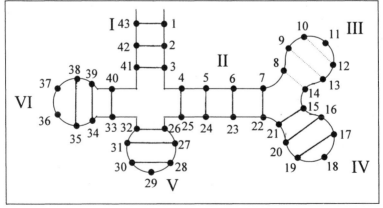

Figure 9.6

The corresponding Nussinov plot:

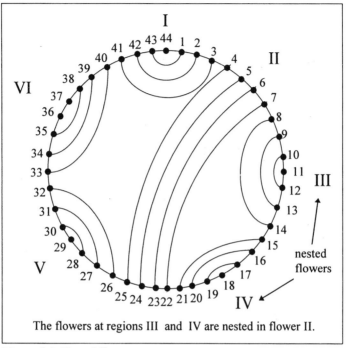

The flowers at regions III and IV are nested in flower II.

Figure 9.7 Nested flowers

How might tertiary structure (three dimensional folding, due to knotted or pseudoknotted structure) appear in a Nussinov plot? Let there be a bond between base 26 and base 39 [30, p. 133] (see Figures 9.8 and 9.9), which otherwise might be unpaired bases [30, p. 152].

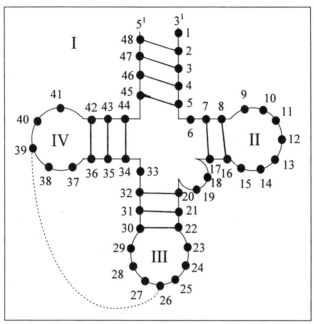

Figure 9.8 Representation of RNA tertiary structure

The corresponding Nussinov plot:

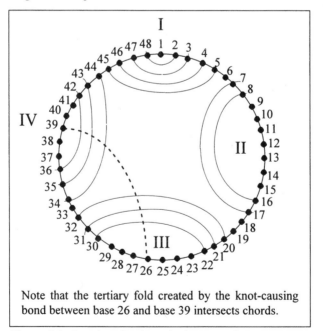

Note that the tertiary fold created by the knot-causing bond between base 26 and base 39 intersects chords.

Figure 9.9 Crossing chords equivalent to tertiary structures

We should note that some of the folding that takes place is intimately involved with base-mismatching; thus these folds are very involved with stability and thermodynamic studies [30, pp. 136-138].

Now that Nussinov plots have been discussed, let us proceed with our consideration of knots and pseudoknots (folding of secondary tRNA into tertiary structures). In this respect, it is important to get some characterization of what a knot or pseudoknot is. Figure 9.10, and the definition have been provided to accomplish this very thing [187, p. 593].

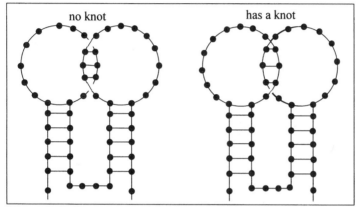

Figure 9.10 Knots vs. pseudoknots

A definition of a secondary structure has been provided. The definition is as follows:

Given that B_t refers to a nucleotide base, then:

$$\left\{ \left(B_i, \ B_j \right) \ | \ 1 \le i \ \text{and} \ i < j \le N \right\} - X$$

where X excludes mismatches and doubly linked bases:

$$\left\{ \left(B_i, \ B_j \right) \ \text{and} \ \left(B_i, \ B_k \right) \ | \ j \ne k \right\}$$

However, we wish to add a strong constraint: no knots.

$$\left\{ \left(B_h, \ B_j \right) \ \text{and} \ \left(B_i, \ B_k \right) \ | \ h < i < j < k \right\}$$

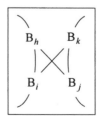

Figure 9.11

Examining Figure 9.11 with knots at the bottom of the previous page very closely, we find:

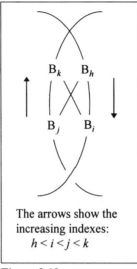

The arrows show the
increasing indexes:
$h < i < j < k$

Figure 9.12

We should take special note that nothing is known about the thermodynamics of knots. As knots are really due to unpaired secondary structure bases participating in tertiary folding, we shall defer further analysis of knots in tertiary structures until we have a better understanding of secondary (planar) structures.

An example showing knotted structure follows. It is for yeast phenylalanine tRNA (specific bases not shown).

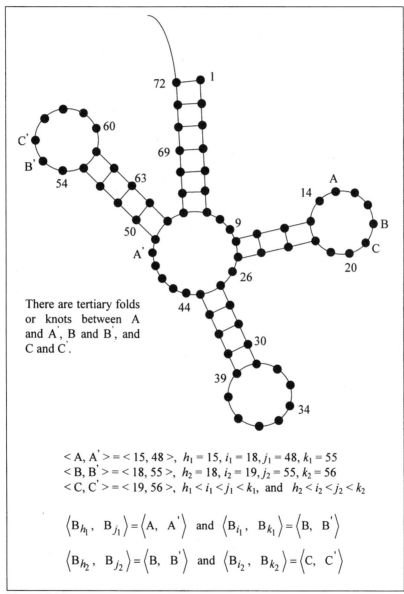

$< A, A' > = < 15, 48 >$, $h_1 = 15$, $i_1 = 18$, $j_1 = 48$, $k_1 = 55$
$< B, B' > = < 18, 55 >$, $h_2 = 18$, $i_2 = 19$, $j_2 = 55$, $k_2 = 56$
$< C, C' > = < 19, 56 >$, $h_1 < i_1 < j_1 < k_1$, and $h_2 < i_2 < j_2 < k_2$

$$\langle B_{h_1}, \ B_{j_1} \rangle = \langle A, \ A' \rangle \ \text{ and } \ \langle B_{i_1}, \ B_{k_1} \rangle = \langle B, \ B' \rangle$$

$$\langle B_{h_2}, \ B_{j_2} \rangle = \langle B, \ B' \rangle \ \text{ and } \ \langle B_{i_2}, \ B_{k_2} \rangle = \langle C, \ C' \rangle$$

Figure 9.13

In order to better classify the decomposition of secondary structures, the following is provided [187, p. 595].

$< B_i, B_j >$ are paired and $i < r < j$, and there is no pair $< B_x, B_y >$ between $< B_i, B_j >$ and B_r; then B_r is accessible from B_i, and B_j:

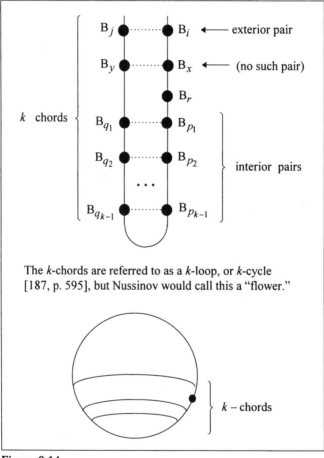

The k-chords are referred to as a k-loop, or k-cycle [187, p. 595], but Nussinov would call this a "flower."

Figure 9.14

Four cases are distinguished; see Figure 9.15.

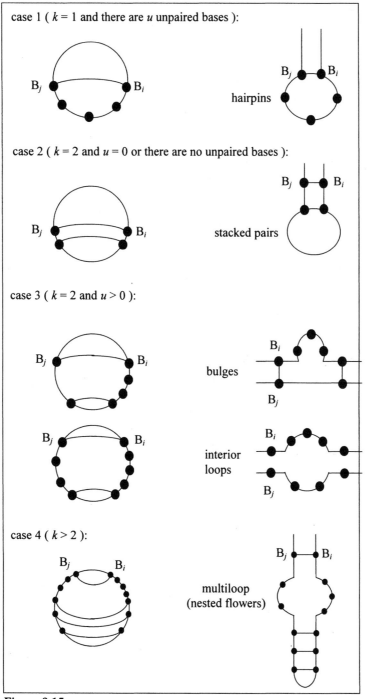

Figure 9.15

Before we proceed further, the bonds between unpaired nucleotide bases, usually at a distance, that were implicated in tertiary folding are what became called the knotted or pseudoknotted structure in RNA. Precision is significant: "The result of such a base pairing was initially called a knotted structure, because of the potential formation of real knots..." and "A general definition of a pseudoknot may be given as follows: a structural element of RNA formed on base pairing of nucleotides within one of the four loops in an orthodox secondary structure ... with nucleotides outside that loop. According to this definition, base pairing is not restricted to interactions between one of the loops and the free single-stranded stretches at either the 3' or 5' terminus of the RNA chain ..." [140, pp. 290, 291]. We shall be even more explicit, "No true knots in RNA molecules however have been reported. We therefore prefer the term 'pseudoknot' ... " [139, p. 1728]. However, the possibility of true knots does exist [139, p. 1728]. The complicated, tertiary folding of RNA predicted on the basis of the sequences of nucleotide bases alone is not yet a possibility [140, p. 300]. However, Pulleyblank [192], when studying DNA topoisomerases, does deal with instances of "catenated" and "knotted" DNA.

A number of recent papers[a][b] and books[c][d] discuss molecular knots in DNA, RNA and proteins (both artificially synthesized as well as naturally occurring molecules). Catenanes, rotaxanes, DNA, RNA[e], and proteins[f][g]. Molecules in the form of the famous non-planar Kurotowski K_5 exist. Möbius ladders of three rungs may be embedded in $K_{3,3}$. Molecules displaying trefoil and Borromean knot structure (and other knots) have been synthesized. Synthetic, stable Holliday form DNA with 4 branches with overhangs on each branch, may be ligated to create 2-dimensional planar networks. Such 2-dimensional structures suggest that shape grammars might be applied. The Appendix contains a discussion of shape grammars with an example that focuses upon Holliday structures, as well as other possible applications.

Kinoplast DNA (kDNA) found in mitochondria of trypanosomes occur in the form of networks of catenated chains or figure eight structures [90, pp 605-606]. These networks are duplicated with the mediation of topoisomers[h]. Trefoil knots and catenated structures do occur naturally in metalloproteins. Catenanes have been found *in vivo* in mouse plasmid DNA. Molecular topology and molecular knots are beginning to become ever more significant, and this certainly applies to RNA, the subject of this chapter.

---

[a] J. S. Siegal, Chemical Topology and Interlocking Molecules, SCIENCE, May 28, 2004, 304, 5675, 1256-1258.

[b] L. Wang, M. O. Vysotsky, A. Bogdan, M. Bolte, and V. Böhmer, Multiple Catenanes Derived from Calix[4]arenes, SCIENCE, May 28, 2004, 304, 5675, 1312-1314.

[c] J. -P. Sauvage, C. Dietrich-Buchecker (Editors), Molecular Catenanes, Rotaxanes and Knots, A Journey Through the World of Molecular Topology, Wiley-VCH, 1999.

[d] G. Schill, Catenanes, Rotaxanes, and Knots, Academic Press, 1971.

[e] N. C. Seeman, H. Wang, R. DiGate, An RNA topoisomerase, Proceedings of the National Academy of Sciences USA, September 1996, 93, 9477-9482.

[f] K. Mislow, C. Liang, Knots in Proteins, Journal of the American Chemical Society, 1994, 116, 24, 11189-11190.

[g] K. Mislow, C. Liang, Knots in Proteins, Journal of the American Chemical Society, 1995, 117, 15, 4201-4213.

[h] N. Cozzarelli, J. Chen, P. Englund, Changes in network topology during replication of kinetoplast DNA, The European Molecular Biology Organization Journal (EMBO), 1995, 14, 24, 6339-6347.

In order to obtain more knowledge of RNA folding, some more examples of knotted structures and pseudoknotted structures such as the following, are instructive [1, p. 3036].

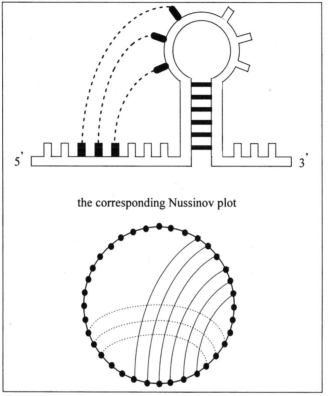

the corresponding Nussinov plot

Figure 9.16 A more complicated Nussinov plot for tertiary folding

The thermodynamic parameters of such folds remain unknown experimentally. Tertiary folding takes place by providing added stability. Thus the rate of such folding is a significant factor too [1, p. 3037]. The dynamic process of RNA folding indicates that these molecules do not constantly reshuffle their secondary structures. Instead, closely knotted short-range interactions of a sequential nature are progressively created, with mostly only long-range bonds being broken and closed again to change the gross structure of the tertiary conformation of RNA [115, pp. 519, 523, 527, 531]. However, new kinds of pseudoknotted structures continue to be found, and one parameter to be considered is the distance between paired bases (distance is proportional to the free energy required to bring portions of RNA into close proximity, and is measured by the number of intervening nucleotide bases [1, pp. 3036, 3038]). Sometimes special base sequences are found in this folding process, such as GGG-CCC triple base pseudoknot folding [139, p. 1719]. For other interesting examples of pseudoknotted folding, see Figures 9.17 and 9.18.

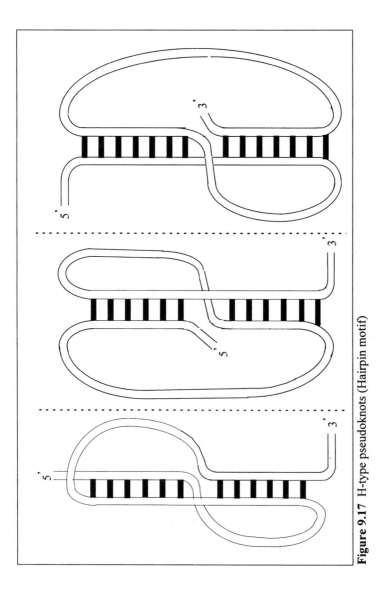

Figure 9.17 H-type pseudoknots (Hairpin motif)

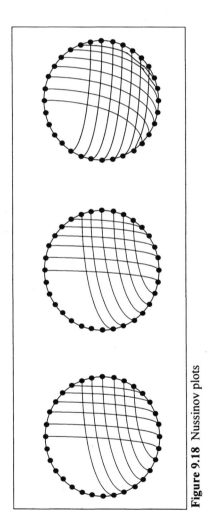

Figure 9.18 Nussinov plots

The above examples of folding were found also [1, p. 3036].

It is important to keep in mind that changes in conformation may increase stability; thus the folding process found Figure 9.19 was also found to be useful [1, p. 3037].

RNA Tertiary Folding

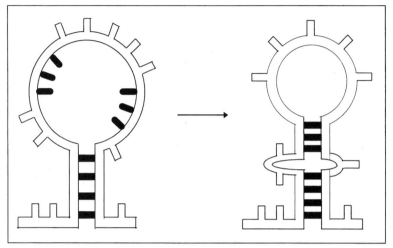

Figure 9.19 Increased tertiary stability changes folding

It is important that the examples of folded structures examined be understood as very simple examples with few bases. Actually studied viruses such as turnip yellow mosaic virus (TYMV) and tobacco mosaic virus (TMV) have hundreds of bases, thus affording many opportunities for complicated folds [1, p. 3042]. Other tRNA structures with complicated tertiary folding, with thousands of nucleotide bases, have been studied as well [139, Table I, p. 1722].

The processes involved in intron/extron splicing in fungal mitochondrial RNA also involve tertiary folding (see the schematic in Figure 9.20). This is especially significant, as the folding is not involved at the 3′ or 5′ terminals, but takes place at internal sites of RNA [139, p. 1726].

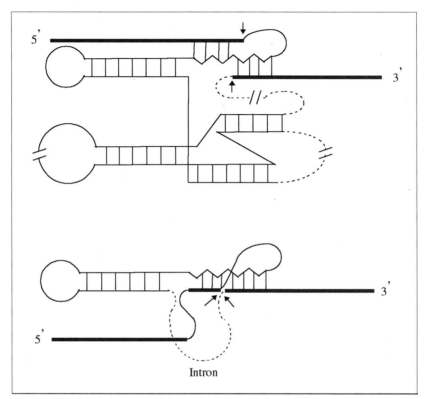

Figure 9.20 Intron/extron splicing and tertiary folding

RNA Dynamic Structural Changes

A novel way of viewing RNA (as well as DNA) dynamic structural changes has been described. A sample RNA structure undergoing dynamic modification (of bonding, due to folding) appears in Figure 9.21. This has already been briefly described in Chapter 3 (parity of nucleotide bases). Bases A, C, G, U/T are assigned complex values i, -1, 1, -i (respectively) [100]. Matrix multiplication may be utilized to support this dynamic modification; an example being found to correspond to the change in Figure 9.21 follows.

The vector of bases associated with this RNA structure.

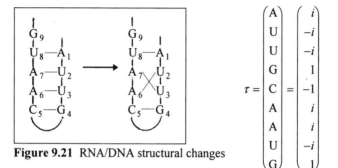

Figure 9.21 RNA/DNA structural changes

$$\tau = \begin{pmatrix} A \\ U \\ U \\ G \\ C \\ A \\ A \\ U \\ G \end{pmatrix} = \begin{pmatrix} i \\ -i \\ -i \\ 1 \\ -1 \\ i \\ i \\ -i \\ 1 \end{pmatrix}$$

$S_1 \tau = S_2 \tau$

thus

Entries in matrix S are easily computed. S is symmetric, thus

1. $s_{i,j} = s_{j,i}$.
2. If base i is paired (bonded) with base j in the secondary structure, then $s_{i,j} = -1$.
3. If base i is not paired with any other base, then $s_{i,i} = -1$ (note, this excludes $s_{i,i}$).
4. $s_{i,i} = 1$ if base i is not paired with any other base; else $s_{i,i} = 0$.

then $\tau = S_1^{-1} S_2 \tau$

but also $S^{-1} = S$ thus $S_1^{-1} S_2 \tau = S_1 S_2 \tau =$

RNA Viewed as Quarternions

The intention is to be able to describe tRNA structures as complicated as the one below.

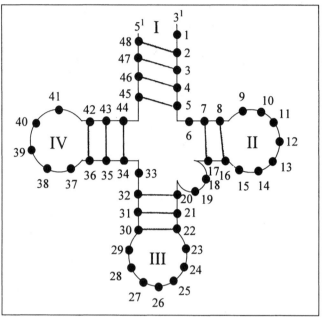

Figure 9.22 Quaternions may be used descriptively

We shall not pursue this any further, except to point out that this analysis may be elaborated in slightly different directions, which will now be discussed. Two directions suggest themselves for further analysis, and both have been pursued. Complex numbers may be viewed as a mathematical group. We shall not study RNA and DNA structures from the point of view of groups, but we will briefly examine the second elaboration: RNA and DNA structures viewed as quaternions [101].

A little notation.: if y represents a nucleotide base chosen from A, C, G, T, or U, then $\overset{c}{y}$ represents the complementary base. This operation of complementarity has already been discussed; thus $y^c = -y$. The viewpoint taken is that in describing dynamic changes that take place during the tertiary folding operation, theorientation of a base must be capable of being described. Thus given nucleotide base X, two orientations are described: $X_\downarrow$ and $X_\uparrow$. With four nucleotide bases (let us restrict ourselves to RNA for the moment), we must be able to describe:

$$A_\downarrow \quad \text{and} \quad A_\uparrow$$
$$C_\downarrow \quad \text{and} \quad C_\uparrow$$
$$G_\downarrow \quad \text{and} \quad G_\uparrow$$
$$U_\downarrow \quad \text{and} \quad U_\uparrow$$

Note: $X_\uparrow = -k\, X_\downarrow$ and $X_\uparrow^c = k\, X_\downarrow$. We may then assign the following values.

$$
\begin{array}{ll}
G_{\downarrow} & 1 \\
C_{\downarrow} & -1 \\
A_{\downarrow} & i \\
U_{\downarrow} & -i
\end{array}
\quad \text{and} \quad
\begin{array}{ll}
G_{\uparrow} & 1 \\
C_{\uparrow} & -1 \\
A_{\uparrow} & i \\
U_{\uparrow} & -i
\end{array}
\quad \text{(an isomorphism)}
$$

Thus there are a total of eight nucleotide "states." Viewed as quarternions, we have the following:

Note:
$$
\begin{aligned}
G_{\uparrow} &= -k \cdot G_{\downarrow} &= -k \cdot 1 = -k \\
C_{\uparrow} &= -k \cdot C_{\downarrow} &= -k \cdot -1 = k \\
A_{\uparrow} &= -k \cdot A_{\downarrow} &= -k \cdot i = -j \\
U_{\uparrow} &= -k \cdot U_{\downarrow} &= -k \cdot -i = j
\end{aligned}
$$

where the usual quaternion relationships hold:
$$
\begin{array}{lll}
k\,i = j, & i\,j = k, & j\,k = i, \\
i\,k = -k\,i, & j\,i = -i\,j, & k\,j = -j\,k, \\
i^2 = j^2 = k^2 = i\,j\,k = -1
\end{array}
$$

As quaternions may be described by 4×4 matrices, we also have the following.

$$
G_{\downarrow},\ G_{\uparrow} = \left(1\right) =
\begin{pmatrix}
1 & 0 & 0 & 0 \\
0 & 1 & 0 & 0 \\
0 & 0 & 1 & 0 \\
0 & 0 & 0 & 1
\end{pmatrix}
\qquad
A_{\downarrow},\ A_{\uparrow} = \left(i\right) =
\begin{pmatrix}
0 & 0 & 0 & 1 \\
1 & 0 & 0 & 0 \\
0 & 1 & 0 & 0 \\
0 & 0 & 1 & 0
\end{pmatrix}
$$

$$
C_{\downarrow},\ C_{\uparrow} = \left(-1\right) =
\begin{pmatrix}
0 & 0 & 1 & 0 \\
0 & 0 & 0 & 1 \\
1 & 0 & 0 & 0 \\
0 & 1 & 0 & 0
\end{pmatrix}
\qquad
U_{\downarrow},\ U_{\uparrow} = \left(-i\right) =
\begin{pmatrix}
0 & 1 & 0 & 0 \\
0 & 0 & 1 & 0 \\
0 & 0 & 0 & 1 \\
1 & 0 & 0 & 0
\end{pmatrix}
$$

These quarternion matrices may be used for parity checking.

We shall end our discussion here. The important conclusion is that rich possibilities exist in applying aspects of the theory of computation to RNA and DNA tertiary structures, and the associated dynamics.

As a closing note, the following quote: "...the analysis of the secondary structure dynamics is very awkward, especially in cases when the dimension of the transition matrix is large" [102, p. 40]. Thus an S matrix of complex numbers when many nucleotide bases are to be described may not be very practical: the more so, if complex numbers are extended to quarternions.

Stochastic Context-Free Languages and RNA

We have closed the discussion of RNA structure analysis using complex numbers and quaternions, but we make brief mention of the structural analysis of RNA using stochastic context-free grammars (sometimes referred to as probabilistic context-free grammars) [105], [106].

A note: RNA may have alternative structures. This is especially true as the accepted view is that the specific conformation of RNA is dependent upon the free energy associated with the conformation, but tertiary conformation is based upon folds due to knots and pseudoknots, while the free energy of knots and pseudoknots is not usually computed. The consequence: alternative RNA structures are hard to deal with mathematically, and in fact RNA has a variety of stable or semistable conformations. Accepting that a given RNA molecule may have a variety of conformations, or that the conformation is ambiguous, a recent approach to RNA structures is to use context-free grammars with parsers that can support ambiguity (multiple parse trees), specifically, the use of stochastic ambiguous context-free grammars. One of the most well-known methods of dealing with ambiguous context-free grammar parsers is to use CYK parsing (a form of dynamic programming). We shall not get into the details of CYK parsing–many books dealing with the subject of parsing cover this. It is sufficient to point out that the CYK parser requires that the grammar be in Chomsky normal form. Every context-free grammar may be put into Chomsky normal form (CNF, see Chapter 2). A context-free grammar is in CNF if every production rule has either of two forms:

$A \Rightarrow a$ or $A \Rightarrow B\,C$, where: A, B, C are non-terminals and "a" is a terminal.

A context-free grammar is said to be stochastic if the following is true for all production rules with non-terminal A on the left.

$A \Rightarrow \alpha_1 \,|\, \alpha_2 \,|\, \ldots \,|\, \alpha_n$ then $A \Rightarrow \alpha_i$ is associated with weight p_i; then if

$$\sum_{i=1}^{n} p_i = 1,$$ the weights form a probability distribution

Combining the two conditions for the grammar:
1. The grammar is in CNF.
2. The grammar is stochastic.

Every production rule in the context-free grammar must be of the following forms only.
1. $A \Rightarrow a$ is the only production rule with non-terminal A on the left, $p_A = 1$
2. $A \Rightarrow BC$ is the only production rule with non-terminal A on the left, $p_A = 1$
3. $A \Rightarrow a \,|\, BC$ are the only two production rules with non-terminal A on the left,
                       then $A \Rightarrow a$ has probability p, $A \Rightarrow BC$ has probability q,
                       and $\quad p + q = 1$

We shall not dwell upon the technicalities of stochastic context-free grammars in a CYK parsing environment any further, but shall view these grammars as simply context-free from here on.

We find the following grammar provided as an example.

$$G = \left(\{S_i\}_{i=0}^{13}, \{A, C, G, T\}, \mathcal{P}, S_0 \right)$$

where the production rules $\mathcal{P}$ are as follows.

$$
\begin{aligned}
S_0 &\Rightarrow S_1 \\
S_1 &\Rightarrow AS_2U \mid CS_2G \\
S_2 &\Rightarrow AS_3U \\
S_3 &\Rightarrow S_4S_9 \\
S_4 &\Rightarrow US_5A \\
S_5 &\Rightarrow CS_6G \\
S_6 &\Rightarrow AS_7 \\
S_7 &\Rightarrow GS_8 \mid US_7 \\
S_8 &\Rightarrow G \mid U \\
S_9 &\Rightarrow AS_{10}U \\
S_{10} &\Rightarrow CS_{10}G \mid GS_{11}C \\
S_{11} &\Rightarrow AS_{12}U \\
S_{12} &\Rightarrow US_{13} \\
S_{13} &\Rightarrow C
\end{aligned}
$$

If any one thing clearly stands out here, it is that this grammar is not in CNF.

The following derivation is provided (modified here); as an example (probabilities omitted), see Figure 9.23.

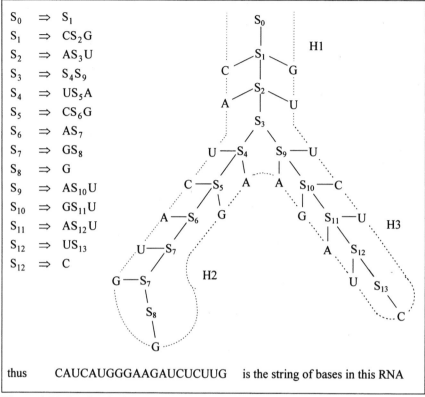

$$
\begin{array}{rcl}
S_0 &\Rightarrow& S_1 \\
S_1 &\Rightarrow& CS_2G \\
S_2 &\Rightarrow& AS_3U \\
S_3 &\Rightarrow& S_4S_9 \\
S_4 &\Rightarrow& US_5A \\
S_5 &\Rightarrow& CS_6G \\
S_6 &\Rightarrow& AS_7 \\
S_7 &\Rightarrow& GS_8 \\
S_8 &\Rightarrow& G \\
S_9 &\Rightarrow& AS_{10}U \\
S_{10} &\Rightarrow& GS_{11}U \\
S_{11} &\Rightarrow& AS_{12}U \\
S_{12} &\Rightarrow& US_{13} \\
S_{12} &\Rightarrow& C
\end{array}
$$

thus   CAUCAUGGGAAGAUCUCUUG   is the string of bases in this RNA

Figure 9.23 Stochastic Context-Free Languages and RNA structure

We can very easily modify this grammar to be in CNF.

A single probability with all of these production rules,
except for two probabilities for S_1, S_7, and S_{10}.

S_0	$\Rightarrow$	S_1	T_1	$\Rightarrow$	S_2U
S_1	$\Rightarrow$	$AT_1 \mid CT_2$	T_2	$\Rightarrow$	S_2G
S_2	$\Rightarrow$	AT_3	T_3	$\Rightarrow$	S_3U
S_3	$\Rightarrow$	S_4S_9	T_4	$\Rightarrow$	S_5A
S_4	$\Rightarrow$	UT_4	T_5	$\Rightarrow$	S_6G
S_5	$\Rightarrow$	CT_5	T_6	$\Rightarrow$	$S_{10}U$
S_6	$\Rightarrow$	AS_7	T_7	$\Rightarrow$	$S_{10}G$
S_7	$\Rightarrow$	$GS_8 \mid US_7$	T_8	$\Rightarrow$	$S_{11}C$
S_8	$\Rightarrow$	$G \mid U$	T_9	$\Rightarrow$	$S_{12}U$
S_9	$\Rightarrow$	AT_6			
S_{10}	$\Rightarrow$	$CT_7 \mid GT_8$			
S_{11}	$\Rightarrow$	AT_9			
S_{12}	$\Rightarrow$	US_{13}			
S_{13}	$\Rightarrow$	C			

When we consider the complicated structures found in secondary as well as tertiary RNA structures, as reviewed in the beginning of this discussion about RNA, we cannot but feel that the approach based upon stochastic context-free grammars has a long way to go to catch up, but it is a beginning!

Along these same lines, the following context-free grammar has also been proposed [160, p. 586]. However, the grammar has been modified for RNA, as opposed to DNA.

$$G = \left(\{S\}, \ \{a, \ c, \ g, \ u\}, \ \mathcal{P}, \ S \right)$$

where: $\mathcal{P} = \left\{ S \Rightarrow SS \ | \ aSu \ | \ cSg \ | \ gSc \ | \ uSa \ | \ \lambda \right\}$

Example of a generation of a RNA-like structure:

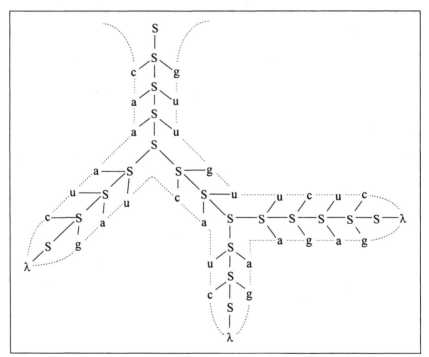

Figure 9.24 Another example of RNA structures that viewed statistically

Chapter 10: Semigroups and Bioinformatics

In England, jails were private property. Prisoners were loaded with chains in order to extract bribes for easement of irons. Typhus epidemics with foci in jails were common, called Black Assizes, circa the 16th and 17th century. Typhus is referred to historically as "morbus carcerorum"–death in incarceration. The name was justified. Thus one aspect of the relationship between medicine and "justice" has been ellucidated. Typhus is often referred to as "jail fever" or "ship fever" [232]. One might also recall that the rotting hulks of ships were commonly used as jails.

Social Views About Tuberculosis

Geography, climate: Southern, tropical environments are worse for TB, as compared to northern, cooler climates [226, p. 144].

"sanitarium"
etymology: Healthful room (or place to do something: maintain health)
"sanitorium"
etymology: Curative or healing room (or place to do something: improve health)

sanitoria: Located at the seashore, high mountain tops, deserts [226, p. 144].

Areas free of mists, fogs, smoke. Areas of low humidity, little extremes in temperature [226, p. 144].

Patients' health may be restored in a sanitarium, and society is protected from contagion (TB patients may be committed involuntarily to a sanitarium) [226, p. 172].

Finite State-Space Model of Metabolism

A multi-enzyme system M is defined as $M = (E, S, I)$, where:

E   is a set of enzymes (proteins), written as E_1, E_2, etc.
S   is a set of substrates, written as s_1, s_2, etc.
I   is a set of inorganic ions

The metabolic state-space is $Q_M = \left\{ q_i \mid q_i \in E \cup I \cup \{s_i\} \text{ and } s_i \in S \right\}$

Given this definition, then state transitions take place due to the action of co-enzymes. The following state-space is provided as a model of a portion of bacterial intermediary metabolism [92].

It is claimed that the mathematical object that corresponds to this 64 state machine is either a permutation group or in general, a semigroup: order between $64!$ and 64^{64}. See Figure 10.1.

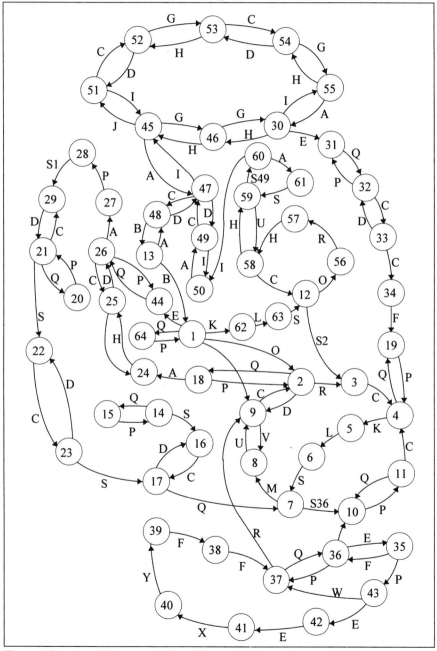

Figure 10.1 A partial representation of bacterial intermediary metabolism

Semigroup for the Krebs Cycle

Biochemical Analysis of Metabolic Pathways

[68, pp. 69-71]

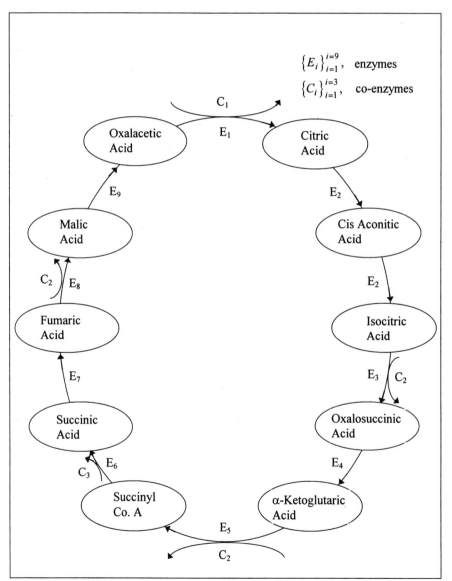

Figure 10.2 Krebs Cycle

The Krebs cycle may be simplified by combining reactions with co-enzymes, and using reaction rates, as follows.

$$\{S_1,\ S_4,\ S_6,\ S_7,\ S_9\},\ \text{Substrates}$$

$$\{C_i\}_{i=1}^{i=3},\qquad\text{Co-Enzymes}$$

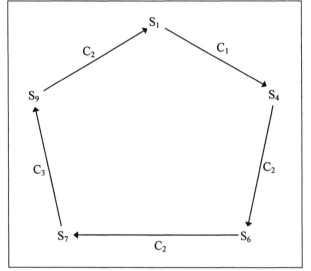

Figure 10.3 Simplified Krebs Cycle

Machine realization:

	C_1	C_2	C_3
S_1	S_4	S_1	S_1
S_4	S_4	S_6	S_4
S_6	S_6	S_7	S_6
S_7	S_7	S_7	S_9
S_9	S_9	S_1	S_9

generators

$$s = \begin{pmatrix} S_1 & S_4 & S_6 & S_7 & S_9 \\ S_4 & S_4 & S_6 & S_7 & S_9 \end{pmatrix}$$

$$y = \begin{pmatrix} S_1 & S_4 & S_6 & S_7 & S_9 \\ S_1 & S_6 & S_7 & S_7 & S_1 \end{pmatrix}$$

$$r = \begin{pmatrix} S_1 & S_4 & S_6 & S_7 & S_9 \\ S_1 & S_4 & S_6 & S_9 & S_9 \end{pmatrix}$$

The semigroup associated with this realization has 183 elements.

Further analysis of this application shall not be studied here, but [166] should be examined. Other applications in the life sciences are studied under the subject of Lindenmeyer $\mathcal{L}$ systems.

Part II

Automata Theory and Disciplines Other than Bioinformatics

The majority of studies concerning automata theory, or mathematical linguistics, or computability are in the traditional areas of computer science, engineering, and mathematics. Now it is hoped that this book has collected evidence that there is a rapidly growing area of bioinformatics. However, the author's objective in this book is not limited to bioinformatics. Applications of automata theory in other diverse fields of study are also important.

The results found in other areas of study are of an uneven quality, often limited. Perhaps mathematical linguistics or computability is not a good model in other subject areas, or (as the author believes) perhaps automata theory is a theory that is not too widely known to researchers in other areas. In any case, I provide in Part II of this book a summary of papers that indicate that computability theory might be useful in studies such as medicine, psychology, anthropology, sociology, geology, medicine, etc.

The papers we shall discuss include:
1. Nursing seal pups
2. Bird songs viewed as a language: the wood peewee
3. Bird songs viewed as a language: the sparrow
4. Bird songs viewed as a language: the black-capped chick-a-dee
5. The descriptive flight of foraging bees
6. Linguistic patterns found in a medical pathology: Tourette's syndrome
7. Applications in geology
8. The use of Linear Sequential Machines in the medical application of ACTH secretion
9. The use of Linear Sequential Machines in analyzing magic
10. The anthropological analysis of the Australian Arunta kinship groups

A few other papers were found, but they emphasized cellular automata (geology) or neural nets–less traditional subjects of automata theory that the author chose not to cover in this book. The author believes that computability theory could be useful in studies in other related areas, such as the following:

> Primates other than human beings
> Cetaceans (whales, porpoises, dolphins)
> Pinnipeds other than seals
> Social insects, other than bees (ants, termites, wasps)
> Geology (vulcanism, earthquake distribution, riverbed formation, etc.)
> Oceanography (deep sea water currents, vortices such as water spouts, whirlpools)
> Meteorology (climate change, wind currents, vortices such as hurricanes,
> cyclones, tornadoes]),
> Music

The author did search the literature, and also contacted research scientists (seeking reference papers), but without success. The future will, however result in more applications of automata theory in such diverse areas of research.

Chapter 11: Automata Theory: Non-Bioinformatics Emergent Computation

"Reports of a more elaborate Chinese method [of smallpox inoculation] involving the insertion of a suitable infected swab of cotton inside the patient's nostril, reached London in 1700. Chinese texts assert that this practice had been introduced into China at the beginning of the eleventh century by a wandering wise man come from the Indian borderland. Subsequently it is said to have become very popular. It therefore seems probable that deliberate inoculation of children with smallpox had been a folk practice in much of Asia for centuries, long before it came to the attention of European doctors and penetrated the repertory of their officially approved techniques in the course of the eighteenth century" [230, p. 224].

"In Turkey, however, smallpox inoculation had been practiced, at least in some milieux, earlier than anywhere else in Europe. It was, in fact, from Turkey that smallpox inoculation reached England, having been introduced to London in 1721, along with other oriental exotica like bloomers and the fez, by Lady Mary Wortley Montagu, wife of a returned ambassador to the Porte. A pair of Greek doctors in Constantinople, who had achieved familiarity with western medicine at the famous medical school of Padua, served as go-betweens. They transmitted information about the folk practice of Turkey to the learned community of Europe by writing a pair of pamphlets on the subject that were widely reproduced in England and elsewhere. According to their report, it was generally believed in Constantinople that the practice of inoculation had long been familiar among Greek peasant women of the Morea and Thessaly" [230, pp. 223, 224].

"...The discovery of vaccination by Edward Jenner, an alert English country doctor, who published his results to the world in 1798. He had noticed that milkmaids seemed never to suffer from smallpox and surmised that they instead contracted cowpox from the animals they tended. Experiments with inoculation of human patients with cowpox showed that immunity to human smallpox did indeed result; and the dangers from cowpox for humans was negligible" [230, p. 222].

"...Catherine the Great introduced inoculation into Russia in 1768 by importing an English doctor to immunize herself and the Crown Prince; but only the court benefited from the Englishman's expertise" [230, p. 223].

"In America, the fearful power of the disease to kill adults was frequently demonstrated by outbreaks among Indians; and the rural and small-town structure of the colonial society ... was also very vulnerable to sporadic epidemics. ... White settlement along the frontier was assisted also by the fact that destruction of Indian populations by infectious diseases, of which smallpox remained the most formidable, continued unabated. The ravages of smallpox among Indians may in fact have been assisted by deliberate efforts at germ warfare. In 1763, for instance, Lord Jeffrey Amherst ordered that blankets infected with smallpox be distributed among enemy tribes, and the order was acted on" [230, p. 222].

Use of Linguistic Methods in Animal Behavior

Another interesting paper attempts to show that linguistic methods might be very useful in describing biological observations, and helping to point out the need to predict further observations. This paper deals with seal pups as well as the wood peewee bird [143].

The authors provide a series of grammars (which have been slightly modified), that they feel are very useful in describing animal behavior. The first set of grammars is for seal pups.

$$G = \left(\{S, \ T, \ U\}, \ \{A, \ B, \ C, \ D\}, \ \mathcal{P}, \ S \right)$$

where: $\mathcal{P}$ is as follows.

$$S \Rightarrow AT \mid BU \mid CU \mid BD$$
$$T \Rightarrow BU \mid CU \mid BD$$
$$U \Rightarrow AT$$

A sample derivation:

$$S \Rightarrow CU \Rightarrow CAT \Rightarrow CABD$$

This is a grammar to model the nursing behavior of newly born seal pups (phoca vitulina), where

A   means that contact is made by the pup, using its head to the mother's head
B   means contact made by the pup, moving its head towards the mother's nipples
C   means vocalization by the pup
D   means feeding by the pup

Thus CABD means that the pup vocalizes, contacts its head with its mother's head, then contacts its head with the mother's nipples, and then feeds.

We have the obviously non-deterministic corresponding FSA:

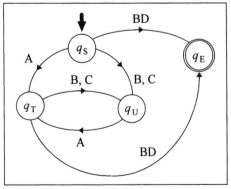

Figure 11.1 FSA for seal pups (phoca vitulina)

There is a second grammar that may be used.

$$G = \left(\left\{S, \ T, \ U, \ V\right\}, \ \left\{A, \ B, \ C, \ D\right\}, \ \mathcal{P}, \ S\right)$$

where: $\mathcal{P}$ is as follows.

$$
\begin{aligned}
S &\Rightarrow AT \mid AU \\
T &\Rightarrow CS \\
U &\Rightarrow BS \mid BV \\
V &\Rightarrow D
\end{aligned}
$$

A sample derivation:

$$S \Rightarrow AU \Rightarrow ABS \Rightarrow ABAU \Rightarrow ABABS \Rightarrow ABABAT \Rightarrow ABABACS$$
$$\Rightarrow ABABACAU \Rightarrow ABABACABV \Rightarrow ABABACABD$$

The FSA that corresponds to this grammar is also not deterministic.

The authors point out that other grammars are possible, which are equivalent with their ability to represent the observed data, such as the following.

$$G = \left(\left\{S, \ U\right\}, \ \left\{A, \ B, \ C, \ D\right\}, \ \mathcal{P}, \ S\right)$$

where: $\mathcal{P}$ is as follows.

$$
\begin{aligned}
S &\Rightarrow ACS \mid AU \\
U &\Rightarrow BS \mid BD
\end{aligned}
$$

Bird Calls

The authors continue by providing us with grammars for the wood pewee (Contopus virens), with a data set of observed bird calls collected by W. Craig over several years of observations.

$$G = \left(\left\{S, \ T, \ U, \ V\right\}, \ \left\{A, \ B, \ C, \ D\right\}, \ \mathcal{P}, \ S\right)$$

where: $\mathcal{P}$ is as follows.

$$
\begin{aligned}
S &\Rightarrow AT \mid CV \\
T &\Rightarrow AT \mid BU \mid CV \\
U &\Rightarrow D \\
V &\Rightarrow AT \mid BU \mid CV
\end{aligned}
$$

A sample derivation:

$$S \Rightarrow CV \Rightarrow CAT \Rightarrow CABU \Rightarrow CABD$$

We are provided with another wood peewee grammar that it is hoped will be more consistent with the data set.

$$G = \left(\{S, \ T, \ U, \ V, \ W, \ X\}, \ \{A, \ B, \ C, \ D\}, \ \mathcal{P}, \ S \right)$$

where: $\mathcal{P}$ is as follows.

$$
\begin{aligned}
S &\Rightarrow AT \mid CV \mid ACW \mid ACABX \\
T &\Rightarrow AT \mid BU \mid ACW \\
U &\Rightarrow D \\
V &\Rightarrow AT \mid BU \mid ACW \mid ACABX \\
W &\Rightarrow AT \mid BU \mid CV \mid ACW \mid ACABX \\
X &\Rightarrow D
\end{aligned}
$$

A sample derivation that the authors say shows that this grammar is ambiguous:

$S \Rightarrow ACABX \Rightarrow ACABD$

$S \Rightarrow ACW \Rightarrow ACAT \Rightarrow ACABU \Rightarrow ACABD$

The authors continue with yet another grammar for this little bird.

$$G = \left(\{S, \ T, \ U, \ V, \ W, \ X\}, \ \{A, \ B, \ C, \ D\}, \ \mathcal{P}, \ S \right)$$

where: $\mathcal{P}$ is as follows.

$$
\begin{aligned}
S &\Rightarrow AT \mid CV \mid ACW \mid AAX \\
T &\Rightarrow BU \mid ACW \\
U &\Rightarrow D \\
V &\Rightarrow AT \mid BU \mid ACW \\
W &\Rightarrow AT \mid BU \mid CV \mid ACW \\
X &\Rightarrow BU
\end{aligned}
$$

However, the authors points out that this grammar might not be entirely adequate, as the terminal sequence CC should be in the language, not the sequence AC; thus we obtain yet another grammar for this tiresome bird!

Thus the authors provide us with yet another grammar.

$$G = \left(\{S,\ T,\ U,\ V,\ W,\ X\},\ \{A,\ B,\ C,\ D\},\ \mathcal{P},\ S\right)$$

where: $\mathcal{P}$ is as follows.

$$
\begin{array}{lcl}
S & \Rightarrow & AT \mid CV \mid AAX \mid CCW \\
T & \Rightarrow & BU \mid CV \mid CCW \\
U & \Rightarrow & D \\
V & \Rightarrow & AT \mid BU \mid AAX \\
W & \Rightarrow & AT \mid BU \\
X & \Rightarrow & BU \mid CV \mid CCW
\end{array}
$$

The authors point out that this grammar is much more consistent with the set of observed bird calls, but that the grammar has a problem. Namely, this grammar is not very efficient, as the right-hand sides for the production rules for T and X are identical.

Thus the authors provides us with another grammar.

$$G = \left(\{S,\ T,\ U,\ V\},\ \{A,\ B,\ C,\ D\},\ \mathcal{P},\ S\right)$$

where: $\mathcal{P}$ is as follows.

$$
\begin{array}{lcl}
S & \Rightarrow & AT \mid AAT \mid CU \mid CCV \\
T & \Rightarrow & CU \mid CCV \mid BD \\
U & \Rightarrow & AT \mid AAT \mid BD \\
V & \Rightarrow & AT \mid BD
\end{array}
$$

The authors are taking a lot of time to explain how the grammar finally arrived at should be consistent with observed data, possible erroneous observations, and possible future observations, and at the same time, produce an efficient grammar. Thus the authors point out that the last grammar is not consistent with one observation: CCAABD, thus the need to modify the grammar yet again.

$$G = \left(\{S,\ T,\ U,\ V\},\ \{A,\ B,\ C,\ D\},\ \mathcal{P},\ S\right)$$

where: $\mathcal{P}$ is as follows.

$$
\begin{array}{lcl}
S & \Rightarrow & AT \mid AAT \mid CV \mid CCV \\
T & \Rightarrow & BU \mid CV \mid CCV \\
U & \Rightarrow & D \\
V & \Rightarrow & AT \mid AAT \mid BU
\end{array}
$$

Note the derivation of CCAABD. $S \Rightarrow CCV \Rightarrow CCAAT \Rightarrow CCAABU \Rightarrow CCAABD$

The authors point out that while regular grammars are in many ways consistent with the wood pewee bird calls, other grammatical structures are possible, namely context-free grammars:

$$G = \Big(\{S, \ T, \ U, \ V, \ W\}, \ \{A, \ B, \ C, \ D\}, \ \mathcal{P}, \ S \Big)$$

where: $\mathcal{P}$ is as follows.

$$
\begin{array}{lll}
S & \Rightarrow & VT \mid WU \\
T & \Rightarrow & WU \mid BD \\
U & \Rightarrow & VT \mid BD \\
V & \Rightarrow & A \mid AA \\
W & \Rightarrow & C \mid CC
\end{array}
$$

The authors finally ends with the grammar below.

$$G = \Big(\{S, \ T, \ U\}, \ \{A, \ B, \ C, \ D\}, \ \mathcal{P}, \ S \Big)$$

where: $\mathcal{P}$ is as follows.

$$
\begin{array}{lll}
S & \Rightarrow & AT \mid AAT \mid CU \mid CCU \\
T & \Rightarrow & CU \mid CCU \mid BD \\
U & \Rightarrow & VT \mid BD
\end{array}
$$

The Savannah Sparrow Sings Regularly

It has been observed that the Passerculus sandwichensis (Savannah sparrow) [188], [189] has 17 different song sequences. These song sequences are composed of sound subsequence elements denoted as A, B, C, D, and E in specific combinations. These song subsequences may be described as follows:

A   introduction
B   post-introduction
C   midsong transition
D   trill
E   terminal buzz

These subsequences are roughly in the range of 5.5 KHz to 10 KHz, and are recorded as sonograms, much as those used to study human linguistics by phoneticians. Sparrow songs recorded in Ontario as well as Nova Scotia have been studied, enabling differences or variations in their songs to be compared linguistically.

The 17 observed song sequences are as follows.

4 elements	5 elements	6 elements	7 elements
ABCD	ABCDE	ABCDCD	ABCDCDE
ABDD	ABADD	ABCDCE	ABDCDDE
ABDE	ABDDE	ABDCDE	
	ABCDD	ABDCDD	
	ABDCD	ABCBDE	
		ABCDDE	
		ABCDAE	

As there are only a small finite number of these song sequences, the entire set constitutes a regular language or a Chomsky type 3 regular language, and an equivalent grammar exists. One approach to the problem of constructing such a grammar is to construct the 17 deterministic FSA $\{M_i\}_{i=1}^{17}$ that corresponds to this set of song sequences and then to construct the composite non-deterministic FSA, then convert this to the deterministic FSA and read off the regular expresion or grammar (appropriately simplified). This will result in the FSA in Figure 11.2.

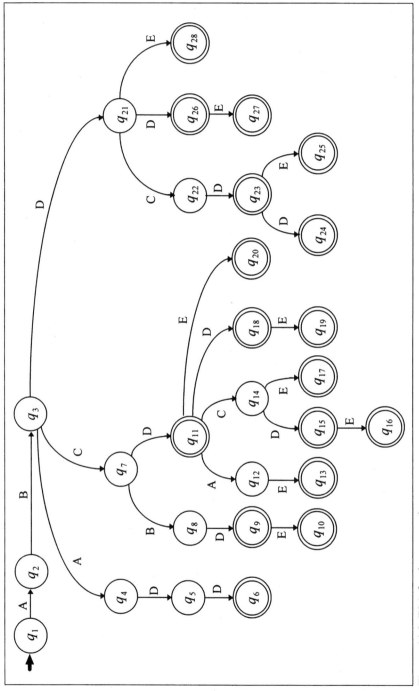

Figure 11.2 FSA for the savannah sparrow

Rather than creating a type 3 grammar that corresponds to this deterministic FSA, we can construct a simplified deterministic FSA, with its correspondingly simplified type 3 grammar.

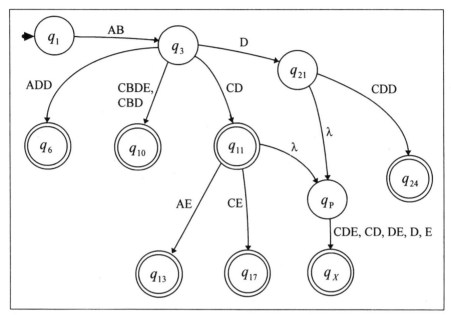

Figure 11.3 FSA for the savannah sparrow

The corresponding type 3 grammar is given by:

$$S_1 \Rightarrow ABS_3$$
$$S_3 \Rightarrow ADD \mid CBDE \mid CBD \mid CDS_{11} \mid DS_{21} \mid CD$$
$$S_{11} \Rightarrow AE \mid CE \mid CDE \mid CD \mid DE \mid D \mid E$$
$$S_{21} \Rightarrow CDD \mid CDE \mid CD \mid DE \mid D \mid E$$

Of course, this is only one possible type 3 grammar, and in fact, as the language contains only 17 (a finite number) words, a grammar isn't even required to specify this language.

Comparison of sparrow songs in Ontario and Nova Scotia has resulted in different FSA given by researchers as follows.

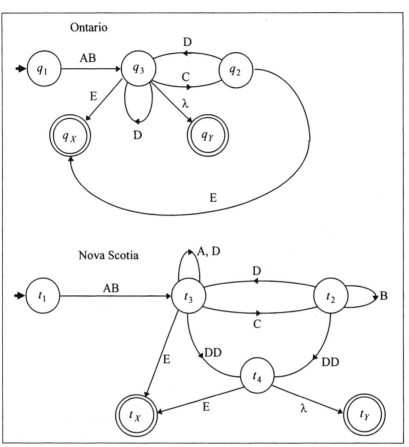

Figure 11.4 FSA for savannah sparrow vs. Nova Scotia sparrow

Given these two FSAs, it is possible to construct variant grammars, but it should be noted that regular expressions are common to both the Ontario and the Nova Scotia sparrows, while there are some Nova Scotia song sequences that the Ontario sparrows do not perform.

ABD*E   is a set of song sequences that sparrows in Ontario and Nova Scotia have in common

ABCBDE   is a Nova Scotia song sequence that has not been observed in Ontario sparrows

Black-Capped Chick-a-dee (Parus atricapillus)

A Turing machine was constructed to describe a language for Parus atricapillus (black-capped chickadee) [193]. In this Turing machine, the read/write head is viewed as stationary, and the tape moves. Modifying the Turing machine so that the read/write head moves, and the tape is stationary, the following Turing machine is described.

	A	B	C	D	#
q_0	q_1, R	q_2, R	q_3, R	q_4, R	q_0, R
q_1	q_1, R	q_2, R	q_3, R	q_4, R	q_7, L
q_2	q_5, R	q_2, R	q_3, R	q_4, R	q_7, L
q_3	q_5, R	q_5, R	q_3, R	q_4, R	q_7, L
q_4	q_5, R	q_5, R	q_5, R	q_4, R	q_7, L
q_5	q_5, R	q_5, R	q_5, R	q_5, R	q_6, L
q_6	q_5, #	q_5, #	q_5, #	q_5, #	q_0, h_{bad}
q_7	q_1, #	q_2, #	q_3, #	q_4, #	q_0, h_{OK}

It is then claimed that the language of Parus atricapillus is as follows.

$L_1 = \left\{ A^a B^b C^c D^d \mid a \geq 0, \ b \geq 0, \ c \geq 0, \ d \geq 0 \right\}$, although rare sequences such as BCB (and others) have been found in the field (young birds, in the process of learning).

In a later paper [194], these bird calls were further analyzed using Markov chains. It is felt that first-order Markov chains are adequate (a call Y only depends upon the previous call X if there is a previous call; thus XY, or Y is all that is required, not subsequences such as WXY, etc.). An attempt is made to provide a semantics or meaning to these calls with different locomotory acts such as flight, landing, twisting on the perch, etc. being a possible semantics. It is pointed out that "Darwin (1871) proposed a theory of sexual selection to explain the evolution of elaborate traits that differ between the sexes, so it is natural to hypothesize that sexual selection has played a part in the evolution of song repertoires..." and "I will first review the evidence that female birds prefer more elaborate repertoires..." (p. 71) [195]. While other semantic interpretations are possible, we needn't subscribe to any specific semantics in this book. It is sufficient that the bird calls are described by a Turing machine. This is significant, as some linguists feel that recursive languages are restricted to human beings, and when confronted with the research in this reference, then go on to elaborate more reasons to ignore scientific findings–their prejudices are firmly rooted. In this later paper, L_1 is redefined as follows.

$$L_2 = \left\{ A^a D^d + B^b C^c D^e + DD + D \mid a \geq 0, \ b \geq 0, \ c \geq 0, \ d \geq 0, \ e \geq 0 \right\}$$

Of course, these languages are effectively equivalent to regular expressions; thus while they may be realized by Turing machines, they may also be realized by pushdown automata or finite state machines.

Foraging Bees

A language to describe the foraging bee is given [184, pp. 145-201]. This language is not claimed to be accurate (it is hypothetical), but it describes a methodology that could be used. I have modified the grammar slightly for convenience. This language is based upon "programmed grammars." A quick definition of "programmed grammars" follows.

We are given production rules of the following form.

$$\alpha \Rightarrow \beta \quad \left\{ \begin{array}{l} \text{successor rules, applied} \\ \text{upon successful} \\ \text{application of the} \\ \text{production rule} \\ \tau \alpha \rho \Rightarrow \tau \beta \rho \end{array} \right\} \quad \left\{ \begin{array}{l} \text{successor rules, applied} \\ \text{upon failed application} \\ \text{of the production rule} \\ \tau \alpha \rho \Rightarrow \tau \beta \rho \end{array} \right\}$$

Note 1: In the successor fields, the null set $\varnothing$ means halt.
Note 2: If a successor field has more than one choice, the choice may be made non-deterministically.

Assumptions will be made about foraging bees.
1. Bees need to collect pollen from about 12 flowers before returning to the hive.
2. Flowers are spaced so that a bee can find a flower in about one meter.
3. If a bee goes from one side of a clump of flowers to the other side, the clumps of flowers are about 5 meters across.
4. The distance between clumps of flowers is about 20 meters.
5. The foraging bee may depart from its original line of flight by $0°$, $45°$, $90°$, or $135°$, (right or left), or the flight may not change at all. These options are symbolized as follows:

$r_d(p)$ right turn by "d" degrees, with probability of "p"
$l_d(p)$ left turn by "d" degrees, with probability of "p"
$s(p)$ no turn, with probability of "p"

Example: $r_{135}(0.2)$ means turn right by $135°$, with probability of 0.2

6. d_x means travel distance x (in meters).
7. h means return to the hive.
8. The current flower and the previous flower determine the foraging bee's angle of flight departure.
9. Each time a foraging bee collects pollen at a flower, a sum of collected pollen is made. When that sum exceeds 12, the bee returns to the hive. If the sum of failures to collect pollen exceed 24, the bee returns to the hive. The following grammar is slightly modified from the one that is discussed. The grammar is as follows:

$$G = \left(\{B, \ M_1, \ M_2, \ M_3, \ M_4 \}, \ V_T, \ \mathcal{P}, \ B \right) \quad \text{where:}$$

$$V_T = \left\{ r_{135}, \ r_{90}, \ r_{45}, \ s, \ l_{45}, \ l_{90}, \ l_{135}, \ d_1, \ d_5, \ d_{20}, \ h \right\}$$

The production rules $\mathcal{P}$:

1	B	$\Rightarrow$	M_1M_2	$\{2(.4),\ 3(.3),\ 4(.3)\}$	$\{\varnothing\}$
2	M_1	$\Rightarrow$	$r_{45}\ d_{20}$	$\{5(R=1),\ 10(R=0)\}$	$\{\varnothing\}$
3	M_1	$\Rightarrow$	$s\ d_{20}$	$\{5(R=1),\ 10(R=0)\}$	$\{\varnothing\}$
4	M_1	$\Rightarrow$	$l_{45}\ d_{20}$	$\{5(R=1),\ 10(R=0)\}$	$\{\varnothing\}$
5	M_2	$\Rightarrow$	M_2M_3	$\{6(.25),\ 7(.25),\ 8(.25),\ 9(.25)\}$	$\{\varnothing\}$
6	M_2	$\Rightarrow$	$r_{90}\ d_1$	$\{14(R=1),\ 19(R=0)\}$	$\{\varnothing\}$
7	M_2	$\Rightarrow$	$r_{45}\ d_1$	$\{14(R=1),\ 19(R=0)\}$	$\{\varnothing\}$
8	M_2	$\Rightarrow$	$l_{45}\ d_1$	$\{14(R=1),\ 19(R=0)\}$	$\{\varnothing\}$
9	M_2	$\Rightarrow$	$l_{90}\ d_1$	$\{14(R=1),\ 19(R=0)\}$	$\{\varnothing\}$
10	M_2	$\Rightarrow$	M_2M_3	$\{11(.4),\ 12(.3),\ 13(.3)\}$	$\{\varnothing\}$
11	M_2	$\Rightarrow$	$r_{45}\ d_5$	$\{25(R=1),\ 31(R=0)\}$	$\{\varnothing\}$
12	M_2	$\Rightarrow$	$s\ d_5$	$\{25(R=1),\ 31(R=0)\}$	$\{\varnothing\}$
13	M_2	$\Rightarrow$	$l_{45}\ d_5$	$\{25(R=1),\ 31(R=0)\}$	$\{\varnothing\}$
14	M_3	$\Rightarrow$	M_4M_3	$\{37 \text{ if } \sum R>12;\ \text{ else } 15(.3),\ 16(.2),\ 17(.2),\ 18(.3)\}$	$\{\varnothing\}$
15	M_4	$\Rightarrow$	$r_{135}\ d_1$	$\{14(R=1),\ 19(R=0)\}$	$\{\varnothing\}$
16	M_4	$\Rightarrow$	$r_{90}\ d_1$	$\{14(R=1),\ 19(R=0)\}$	$\{\varnothing\}$
17	M_4	$\Rightarrow$	$l_{90}\ d_1$	$\{14(R=1),\ 19(R=0)\}$	$\{\varnothing\}$
18	M_4	$\Rightarrow$	$l_{135}\ d_1$	$\{14(R=1),\ 19(R=0)\}$	$\{\varnothing\}$
19	M_3	$\Rightarrow$	M_3M_4	$\{37 \text{ if } \sum R>24;\ \text{ else } 15(.3),\ 20(.2),\ 21(.2),\ 22(.2),\ 23(.2),\ 24(.2)\}$	$\{\varnothing\}$

20	M_3	$\Uparrow$	$r_{90}\,d_5$	$\{25(R=1),\ 31(R=0)\}$	$\{\varnothing\}$
21	M_3	$\Uparrow$	$r_{45}\,d_5$	$\{25(R=1),\ 31(R=0)\}$	$\{\varnothing\}$
22	M_3	$\Uparrow$	$s\ d_5$	$\{25(R=1),\ 31(R=0)\}$	$\{\varnothing\}$
23	M_3	$\Uparrow$	$l_{45}\,d_5$	$\{25(R=1),\ 31(R=0)\}$	$\{\varnothing\}$
24	M_3	$\Uparrow$	$l_{90}\,d_5$	$\{25(R=1),\ 31(R=0)\}$	$\{\varnothing\}$
25	M_3	$\Uparrow$	M_3M_4	$\{37\ \text{if}\ \sum R>12;\ \text{else}\ 26(.3),\ 27(.15),\ 28(.1),\ 29(.15),\ 30(.3)\}$	$\{\varnothing\}$
26	M_3	$\Uparrow$	$r_{90}\,d_1$	$\{14(R=1),\ 19(R=0)\}$	$\{\varnothing\}$
27	M_3	$\Uparrow$	$r_{45}\,d_1$	$\{14(R=1),\ 19(R=0)\}$	$\{\varnothing\}$
28	M_3	$\Uparrow$	$s\ d_1$	$\{14(R=1),\ 19(R=0)\}$	$\{\varnothing\}$
29	M_3	$\Uparrow$	$l_{45}\,d_1$	$\{14(R=1),\ 19(R=0)\}$	$\{\varnothing\}$
30	M_3	$\Uparrow$	$l_{90}\,d_1$	$\{14(R=1),\ 19(R=0)\}$	$\{\varnothing\}$
31	M_3	$\Uparrow$	M_3M_4	$\{37\ \text{if}\ \sum F>24;\ \text{else}\ 32(.2),\ 33(.2),\ 34(.2),\ 35(.2),\ 36(.2)\}$	$\{\varnothing\}$
32	M_3	$\Uparrow$	$r_{90}\,d_{20}$	$\{25(R=1),\ 31(R=0)\}$	$\{\varnothing\}$
33	M_3	$\Uparrow$	$r_{45}\,d_{20}$	$\{25(R=1),\ 31(R=0)\}$	$\{\varnothing\}$
34	M_3	$\Uparrow$	$s\ d_{20}$	$\{25(R=1),\ 31(R=0)\}$	$\{\varnothing\}$
35	M_3	$\Uparrow$	$l_{45}\,d_{20}$	$\{25(R=1),\ 31(R=0)\}$	$\{\varnothing\}$
36	M_3	$\Uparrow$	$l_{90}\,d_{20}$	$\{25(R=1),\ 31(R=0)\}$	$\{\varnothing\}$
37	M_4	$\Uparrow$	h	$\{\varnothing\}$	$\{\varnothing\}$

Thus for example,

$$B \quad \Rightarrow \quad M_1 M_2 \quad \{2(.4), \ 3(.3), \ 4(.3)\} \quad \{\varnothing\}$$

means

apply production rule $B \quad \Rightarrow \quad M_1 M_2$, and
upon success,
if production rule 2 is chosen, then in $M_1 \Rightarrow r_{45} d_{20}$,
r_{45} is $r_{45}(.4)$
if production rule 3 is chosen, then in $M_1 \Rightarrow s \ d_{20}$,
s is $s(.3)$
if production rule 4 is chosen, then in $M_1 \Rightarrow l_{45} d_{20}$;
l_{45} is $l_{45}(.3)$
else upon failure, halt.

Now that the production rules have been clearly explained, with an example as well, let us try a sample derivation in this grammar to explain a bee forage sortie.

Rule #	Production		Choice	d	ΣR	ΣF
1	B	$\Rightarrow M_1 M_2$	$2(.4)$	0	0	0
2	M_1	$\Rightarrow r_{45}(.4) d_{20}$	$5(R=1)$	20	1	0
5	M_2	$\Rightarrow M_2 M_3$	$9(.25)$			
9	M_2	$\Rightarrow l_{90}(.25) d_1$	$14(R=1)$	1	2	0
14	M_3	$\Rightarrow M_4 M_3$	$15(.3)$			
15	M_4	$\Rightarrow r_{135}(.3) d_1$	$14(R=1)$	1	3	0
14	M_3	$\Rightarrow M_4 M_3$	$17(.2)$			
17	M_4	$\Rightarrow l_{90}(.2) d_1$	$14(R=1)$	1	4	0
14	M_3	$\Rightarrow M_4 M_3$	$16(.2)$			
16	M_4	$\Rightarrow r_{90}(.2) d_1$	$14(R=1)$	1	5	0
14	M_3	$\Rightarrow M_4 M_3$	$18(.3)$			
18	M_4	$\Rightarrow l_{135}(.3) d_1$	$14(R=1)$	1	6	0

Rule #	Production			Choice	d	ΣR	ΣF
14	M_3	$\Rightarrow$	$M_4 M_3$	$15(.3)$			
15	M_4	$\Rightarrow$	$r_{135}(.3) d_1$	$14(R = 1)$	1	7	0
14	M_3	$\Rightarrow$	$M_4 M_3$	$17(.2)$			
17	M_4	$\Rightarrow$	$l_{90}(.2) d_1$	$19(R = 0)$	1	7	1
19	M_3	$\Rightarrow$	$M_3 M_4$	$15(.3)$			
15	M_4	$\Rightarrow$	$r_{135}(.3) d_1$	$14(R = 1)$	1	8	1
14	M_3	$\Rightarrow$	$M_4 M_3$	$19(R = 0)$		8	2
19	M_3	$\Rightarrow$	$M_3 M_4$	$15(.3)$			
15	M_4	$\Rightarrow$	$r_{135}(.3) d_1$	$14(R = 1)$	1	9	2
14	M_3	$\Rightarrow$	$M_4 M_3$	$16(.2)$			
16	M_4	$\Rightarrow$	$r_{90}(.2) d_1$	$14(R = 1)$	1	10	2
14	M_3	$\Rightarrow$	$M_4 M_3$	$15(.3)$			
15	M_4	$\Rightarrow$	$r_{135}(.3) d_1$	$14(R = 1)$	1	11	2
14	M_3	$\Rightarrow$	$M_4 M_3$	$18(.3)$			
18	M_4	$\Rightarrow$	$l_{135}(.3) d_1$	$14(R = 1)$	1	12	2
14	M_3	$\Rightarrow$	$M_4 M_3$	$17(.2)$			
17	M_4	$\Rightarrow$	$l_{90}(.2) d_1$	$14(R = 1)$	1	13	2
14	M_3	$\Rightarrow$	$M_4 M_3$	37			
37	M_4	$\Rightarrow$	h				

(continued)

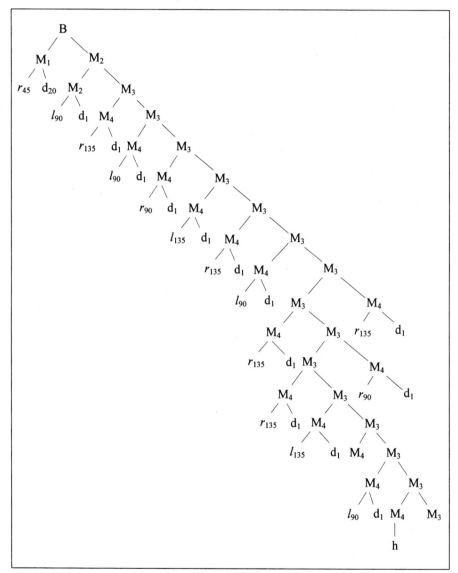

Figure 11.5 Example of foraging sortie in the language of bees

The above language is intended to show how a description of the foraging activity of bees can be described. However, bees also have the linguistic capability to communicate through a "waggle dance." Bees communicate where food has been discovered. The waggle dance consists of an angle from the sun's azimuth that conveys directional information. The waggle dance consists of a figure eight of alternating half circles, the number of figure eights corresponding to a distance. Thus bees communicate the location of food using polar coordinates! In addition, the type of food is also identified. The figure eight/polar coordinates look like the following [214, pp. 532-538], [206, pp. 111-131].

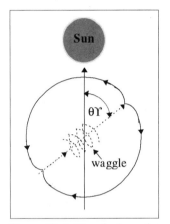

Figure 11.6 Waggle dance language of bees

There was a controversy: did bees really communicate using such a language, or was the dance a mere artifact, and did bees follow a scent to the food? An ingenious experiment was performed, the conclusion being that bees do indeed use a language to communicate. Although we shall not do so, obviously a grammar could be constructed for the effectively non-finite waggle dance language of bees.

Medical Pathology: Tourette's Syndrome

"Psychodynamic theorists would predict that the verbal tics characterizing Gilles de la Tourette's syndrome should exhibit primitive syntax but complex semantics, while neurological considerations lead to an opposite expectation" [104, p. 266].

"In spite of the recent increase of interest in Gilles de la Tourette's disease and the fact that it was first described by Itard over a century ago, there exists no detailed linguistic description of the most salient feature of the syndrome, its verbal tics. Previous discussions of the syndrome have confined themselves to giving a few examples of patients' tics. But consideration has been given to the possibility that the tics emitted by a given patient might follow any sort of internal grammar. The only linguistic formulations concerning the tic are the frequent observations that it is stereotyped in form, draws on a limited 'lexicon', and resembles an interjection. The purpose of this paper is to present a description of the tics exhibited by a patient suffering from the disease and to attempt to formulate a grammar to describe them.

"Gilles de la Tourette's syndrome is characterized by multiple muscular tics, vocal tics which are often coprolalic, and in about twenty per cent of cases by echolalia. Onset is usually during late childhood and the symptoms exhibit a fairly regular course. The muscular tics tend to begin around the eyes as twitches and come to involve progressively the trunk and finally the extremities. The vocal tic usually begins as a inarticulate 'bark' or grunt and only later assumes a verbal form. Symptoms abate during sleep and during sexual arousal...The disease does not seem to have a traumatic or infectious origin, but an organic basis seems likely" [104, p. 266].

The author thus describes Tourette's syndrome very well. The author feels that finite state grammars are suggested as when referring to a particular patient's 11 verbal tics, "...this simplicity of structure, along with the fact that the tics draw from a limited lexicon suggests that the exploration of finite state models of the tic might be profitable" [104, p. 270].

It is significant that while many theorists seek to understand human languages as normally spoken, linguistics may also be important for medical reasons in pathological situations. We will now examine linguistic analysis as applied to Tourette's Syndrome.

The 11 tics referred to by the author will be symbolized as follows.

	Tic
1.	C
2.	FC
3.	FMC
4.	F
5.	FMYC
6.	FMYYC
7.	FRYC
8.	FMYYYC
9.	FRYYC
10.	YYC
11.	FU

Tics are classified in the following manner.

0-limited   (tic elements are independent)
1-limited   (every word in a tic is contextually constrained by the preceding word)
2-limited
etc.

A few FSA are constructed, some being probabilistic; they are like the following.

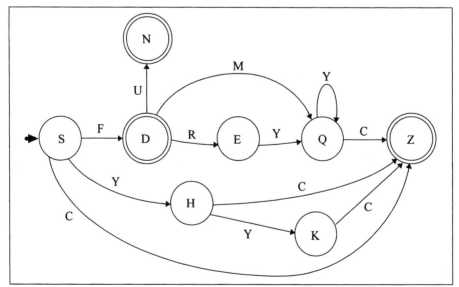

Figure 11.7 FSA for Tourette's syndrome

Geology, Meteorology, Oceanography

Although research did not result in the location of any papers using such techniques as state machines in applications such as geology, meteorology, or oceanography (other than cellular automata and neural nets), such applications might well appear in the future. For example, an admittedly simplistic automaton used in earthquake studies might look something like Figure 11.8.

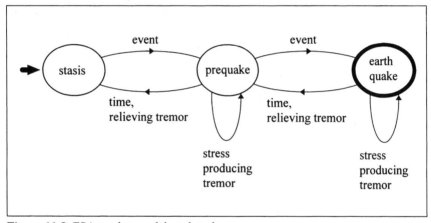

Figure 11.8 FSA used to model earthquakes

Obviously, greater refinement is possible, such as multiple "prequake" states, and a variety of events, and kinds of stress-producing or relieving tremors, etc. Using such an FSA, it is thus possible to create "languages" (type 3, or even more complex) that describe earthquakes.

Similar models could easily be created to account for the shifting of water currents, the melting of polar ice, volcanism, tides, hurricanes, etc.

Closed Loop Cortisol Secretion

An interesting medical application of linear sequential machine theory is cortisol secretion. A few models are given, the simplest one uses feedback and is the one below [54].

Abbreviations	
Stim	Stimulation
RAS	Reticular activating system
ACTH	Adrenocorticotropic hormone
CRF	Corticotropin release factor
PIT	Pituitary gland
ADR	Adrenal gland
DIST	Distribution
Free F	Non-protein bound cortisol
Bound F	Protein bound cortisol

$$y_1 = \text{Stim}$$
$$y_2 = y_1^1$$
$$\text{CRF} = y_2^1 - y_5^1$$
$$y_3 = \text{CRF}^1$$
$$\text{ACTH} = y_3^1$$
$$y_4 = \text{ACTH}^1 - y_4^1$$
$$y_5 = y_4^1$$
$$\text{Bound F} = y_5^1$$

or

$$
\begin{pmatrix} y_1 \\ y_2 \\ CRF \\ y_3 \\ ACTH \\ y_4 \\ y_5 \end{pmatrix} = \begin{pmatrix} 0 & 0 & 0 & 0 & 0 & 0 & 0 \\ 1 & 0 & 0 & 0 & 0 & 0 & 0 \\ 0 & 1 & 0 & 0 & 0 & 0 & -1 \\ 0 & 0 & 1 & 0 & 0 & 0 & 0 \\ 0 & 0 & 0 & 1 & 0 & 0 & 0 \\ 0 & 0 & 0 & 0 & 1 & -1 & 0 \\ 0 & 0 & 0 & 0 & 0 & 1 & 0 \end{pmatrix} \cdot \begin{pmatrix} y_1 \\ y_2 \\ CRF \\ y_3 \\ ACTH \\ y_4 \\ y_5 \end{pmatrix}^1 + \begin{pmatrix} 1 \\ 0 \\ 0 \\ 0 \\ 0 \\ 0 \\ 0 \end{pmatrix} \cdot Stim
$$

$$
Bound\ F\ =\ \begin{pmatrix} 0 & 0 & 0 & 0 & 0 & 0 & 1 \end{pmatrix} \cdot \begin{pmatrix} y_1 \\ y_2 \\ CRF \\ y_3 \\ ACTH \\ y_4 \\ y_5 \end{pmatrix}^1
$$

Thus the LSM is a Moore model, and the transfer function (gain), as well as the sensitivity and stability, of this proposed system could be studied. Our analysis will stop here, although the associated semigroup for this LSM could also be constructed. Examine the LSM given in Fgure 11.9.

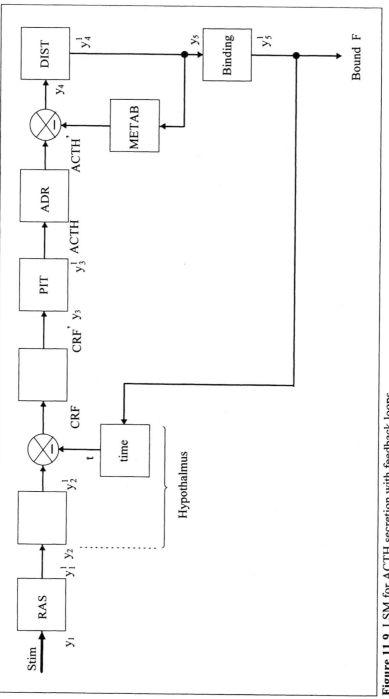

Figure 11.9 LSM for ACTH secretion with feedback loops

An Anthropologist's View of Magic

An interpretation of the anthropologist Malinowski's view of a function of magic [52] has been made in terms of linear systems. Perhaps such a view is valid; some might deny this. Instead of debating, let us present the view.

Let   M   symbolize the amount of magic used
        A   symbolizes the collective anxiety
        U   symbolizes the environmental uncertainty

It is claimed that
        $A \propto U$ (anxiety is proportional to environmental uncertainty)
        $M \propto A$ (magic is proportional to anxiety)
        $A \propto 1/M$ (anxiety is inversely proportional to magic: magic reduces anxiety)

It is further claimed that magic may be modeled as a continuous variable, such that the change in magic over time is proportional to anxiety:

$$\frac{d\,M}{dt} = c_1 A$$

Also, anxiety (viewed as psychological feelings, emotions, or thoughts) may be modeled as a continuous variable, such that the change in anxiety over time is proportional to environmental uncertainty diminished by the imagined efficacy of magic:

$$\frac{dA}{dt} = c_3 U - c_2 M$$

Collecting these equations, we obtain:

$$\frac{d}{dt}\begin{pmatrix} M \\ A \end{pmatrix} = \begin{pmatrix} 0 & c_1 \\ -c_2 & 0 \end{pmatrix}\begin{pmatrix} M \\ A \end{pmatrix} + \begin{pmatrix} 0 \\ c_3 \end{pmatrix} U$$

somewhat more suggestively:

$$\begin{pmatrix} M \\ A \end{pmatrix}^1 = \begin{pmatrix} 0 & c_1 \\ -c_2 & 0 \end{pmatrix}\begin{pmatrix} M \\ A \end{pmatrix} + \begin{pmatrix} 0 \\ c_3 \end{pmatrix} U$$

As is standard practice, we assume that at equilibrium, $\begin{pmatrix} M \\ A \end{pmatrix}^1 = \begin{pmatrix} 0 \\ 0 \end{pmatrix}$,

and we obtain:

$$\begin{pmatrix} 0 \\ 0 \end{pmatrix} = \begin{pmatrix} 0 & c_1 \\ -c_2 & 0 \end{pmatrix}\begin{pmatrix} M \\ A \end{pmatrix} + \begin{pmatrix} 0 \\ c_3 \end{pmatrix} U \quad \text{or} \quad \begin{pmatrix} M \\ A \end{pmatrix} = -\begin{pmatrix} 0 & c_1 \\ -c_2 & 0 \end{pmatrix}^{-1}\begin{pmatrix} 0 \\ c_3 \end{pmatrix} U$$

Thus

$$\begin{pmatrix} M \\ A \end{pmatrix} = -\begin{pmatrix} 0 & \dfrac{1}{c_2} \\ -\dfrac{1}{c_1} & 0 \end{pmatrix}\begin{pmatrix} 0 \\ c_3 \end{pmatrix} U = \begin{pmatrix} \dfrac{c_3}{c_2} \\ 0 \end{pmatrix} U$$

Thus $\quad M = \dfrac{c_3}{c_2} \cdot U \quad$ and $\quad A = 0$

which means that at equilibrium, anxiety is assuaged, and the amount of magic is linearly proportional to environmental uncertainty.

Furthermore, examining the stability of this system by seeking eigenvalues, we find:

$$\begin{pmatrix} 0 & c_1 \\ -c_2 & 0 \end{pmatrix} = \lambda I \quad \text{or} \quad \begin{vmatrix} -\lambda & c_1 \\ -c_2 & -\lambda \end{vmatrix} = 0 \quad \text{thus} \quad \lambda^2 = -c_1 c_2 \quad \text{or} \quad \lambda = \pm i\sqrt{c_1 c_2}$$

The imaginary eigenvalues mean that "the state of the system will not move away from the equilibrium...the system consists in oscillation around equilibrium."

The Algebra of Group Kinship

Most studies in anthropology are not very oriented towards mathematics or mathematical linguistics—but not all studies. Kinship among the Arunta of Australia has been studied from the viewpoint of groups [23].

Functional Notation: φ: $X \to Y$ shall be written as $X\varphi = Y$.

X F Y   means there is a father of X in Y
X M Y   means there is a mother of X in Y
X C = Y means there is a man in X, whose child is in Y
X W = Y means there is a man in X married to a woman in Y

Note: if X F Y, then $Y F^{-1} = X$
        if X C Y, then $Y C^{-1} = X$
        if X W Y, then $Y W^{-1} = X$

also   F = C^{-1} and $M = C^{-1} W$
       C = F^{-1} and $W = F^{-1} M$

For the Arunta, $F = F^{-1}$ thus $F: X \to Y$ and $F^{-1}: Y \to X$; thus

$$X \to Y \text{ and } Y \to X, \text{ or } X \leftrightarrow Y$$

We will use ⎯⎯⎯→ for M

We will use ·········► for F

Then we obtain the following.

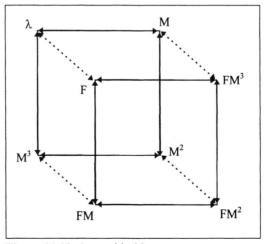

Figure 11.10 Arunta kinship group

We obtain the following Arunta relationship table, a mathematical group.

	λ	M	M^2	M^3	F	FM	FM^2	FM^3
λ	λ	M	M^2	M^3	F	FM	FM^2	FM^3
M	M	M^2	M^3	λ	FM^3	F	FM	FM^2
M^2	M^2	M^3	λ	M	FM^2	FM^3	F	FM
M^3	M^3	λ	M	M^2	FM	FM^2	FM^3	F
F	F	FM	FM^2	FM^3	λ	M	M^2	M^3
FM	FM	FM^2	FM^3	F	M^3	λ	M	M^2
FM^2	FM^2	FM^3	F	FM	M^2	M^3	λ	M
FM^3	FM^3	F	FM	FM^2	M	M^2	M^3	λ

Figure 11.11 Arunta kinship non-Abelian group

Note: This group is not commutative (example: $FM \cdot M = FM^2$, but $M \cdot FM = F$).

The inverses are obvious:

$$\lambda^{-1} = \lambda$$
$$M^{-1} = M^3$$
$$M^{-2} = M^2$$
$$M^{-3} = M^3$$
$$F^{-1} = F$$
$$(FM)^{-1} = FM$$
$$(FM^2)^{-1} = FM^2$$
$$(FM^3)^{-1} = FM^3$$

Some obvious subgroups (we shall omit their cosets):

$$H_0 = \{ \lambda, M^2 \}$$
$$H_1 = \{ \lambda, F \}$$
$$H_2 = \{ \lambda, FM \}$$
$$H_3 = \{ \lambda, FM^2 \}$$
$$H_4 = \{ \lambda, FM^3 \}$$
$$H_5 = \{ \lambda, M, M^2, M^3 \}$$

Boyd investigates other relationships besides F and M, however, the subgrups might also be of interest. We shall stop here.

Appendix[a]

Shape Grammars

In Chapter 9 there is a reference to possible applications of shape grammars. One such application concerns artificial biological materials constructed from a finite number of specially created (stable) Holliday type molecules[b]. Shape grammars might be of practical value in a number of other applications as well.

A shape grammar is defined as follows. $SG = \left(V_M, \ V_T, \ \mathcal{R}, \ I \right)$, where

V_T is a finite set of terminal symbols

V_M is a finite set of marker symbols (that act like non-terminals)

$I \in V_T^* \ V_M^+ \ V_T^*$ is the start symbol.

The production rules of the shape grammar $\mathcal{R}$ are phrase structured, and the rules may be applied serially, or in parallel (the same as with Lindenmeyer systems).

$\mathcal{R}$ is composed of rules of the form $u \Rightarrow v$, where u is a mix of V_T^* with V_M^+, and v is a mix of V_T^* with V_M^*. Objects in V_T and V_M may have any location, orientation, or scaling (think in terms of location in a Cartesian grid or some other coordinate system, rotation, and size), in two or more dimensions. As an example,

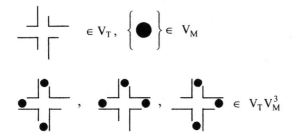

The marker disks may be at any one of the four branches, and marks an active area. In the example that follows, these marked active areas are where ligases may act.

In the following rule for $\mathcal{R}$, the four cruciform symbols on the right of the production rule are of the same size, none are rotated, and all four have the locations indicated.

---

[a] J. Gips, *Shape Grammars and Their Uses: Artificial Perception, Shape Generation and Computer Aesthetics*, 1975, Birkhäuser.

[b] N. C. Seeman, J. Chen, S. M. Du, J. E. Mueller, Y. Zhang, T.-J. Fu, Y. Wang, H. Wang, S. Zhang, Synthetic DNA knots and catenanes, New Journal of Chemistry, 1993, 17, 739-755.

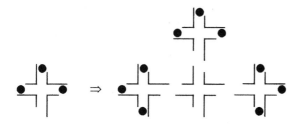

In the following production rule in $\mathcal{R}$, the symbol on the right at the top is larger, has been rotated 45°, and its location (or position) is raised (and the markers have changed, too).

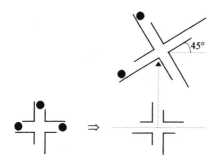

In the shape grammar example to be given, each cruciform symbol signifies a stable Holliday structure with overhangs (blunt ends might work too). These stable Holliday structures may be identical in terms of their base orderings, or alternatively, these cruciferous forms may vary in the ordering of their bases (as long as ligations can occur). In addition, the bases can be naturally occurring or may be artificial bases as discussed in Chapter 3. Note, even if ligases do not yet exist that will ligate artificial bases, the artificial bases need not appear within the overhangs (or within the active areas of the blunt ends). The shape language for Holliday structures is composed of planar (2-dimensional) networks of ligated Holliday structures. It is of no significance that such planar structures occur naturally as artificial structures of this sort might be found to be useful as new bioengineering materials.

An actual example of a shape grammar for Holliday structures follows.

$$SG_{HOL} = \left(V_M, \ V_T, \ \mathcal{R}, \ I \right)$$

L = { all 2-dimensional planar networks of ligated Holliday structures }

Objects of interest on the left hand sides of the rules in $\mathcal{R}$ are in

$$V_T V_M^4 \ \cup \ V_T V_M^3 \ \cup \ V_T V_M^2 \ \cup \ V_T V_M^1$$

In $\binom{4}{n}$ "n" branches of the 4-way Holliday structure may be ligated.

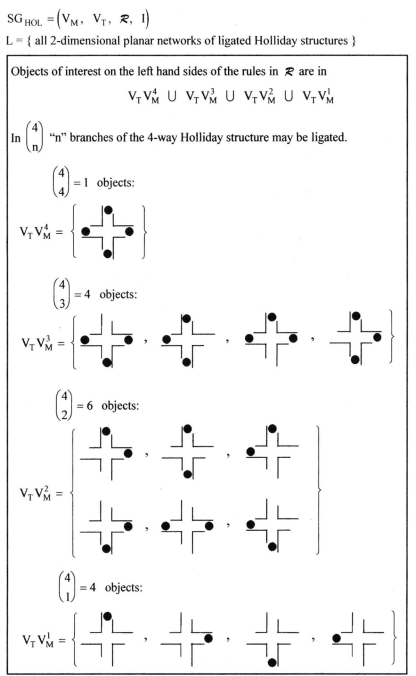

$\binom{4}{4} = 1$ objects:

$$V_T V_M^4 = \left\{ \ \right\}$$

$\binom{4}{3} = 4$ objects:

$$V_T V_M^3 = \left\{ \quad , \quad , \quad , \quad \right\}$$

$\binom{4}{2} = 6$ objects:

$$V_T V_M^2 = \left\{ \quad , \quad , \quad , \quad , \quad , \quad \right\}$$

$\binom{4}{1} = 4$ objects:

$$V_T V_M^1 = \left\{ \quad , \quad , \quad , \quad \right\}$$

Figure A.1 All the different marked objects used for Holliday shape grammar

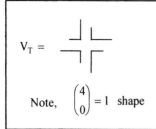

$V_T =$

Note, $\binom{4}{0} = 1$ shape

Note that while a single shape of *stable* Holliday structure is represented here, in fact, a class of such molecules is possible, with variable base orderings, and different base compositions, including "artificial" bases, abasic pairs, etc. as in Chapter 3. The only requirement is that they can be ligated. Which element from V_T is selected can be associated with a probability.

Figure A.2 The single object in V_T for Holliday shape grammar

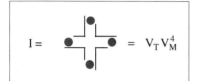

$I =$ $= V_T V_M^4$

Figure A.3 The start symbol for Holliday shape grammar

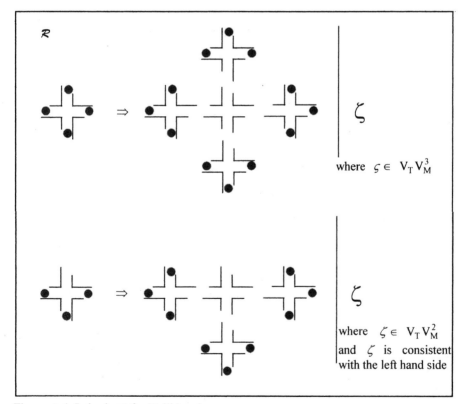

$\mathcal{R}$

ζ

where $\varsigma \in V_T V_M^3$

ζ

where $\zeta \in V_T V_M^2$
and ζ is consistent with the left hand side

Figure A.4 Rules in $\mathcal{R}$ for Holliday shape grammar

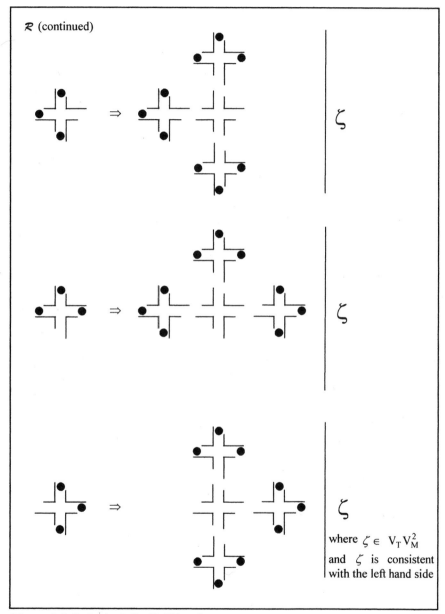

Figure A.5 Rules in $\mathcal{R}$ for Holliday shape grammar

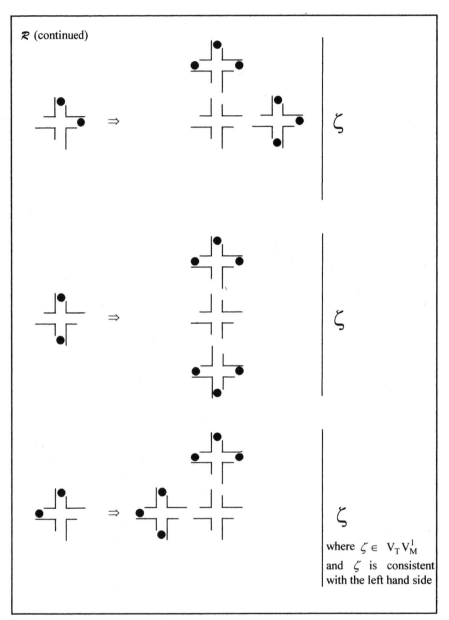

Figure A.6 Rules in $\mathcal{R}$ for Holliday shape grammar

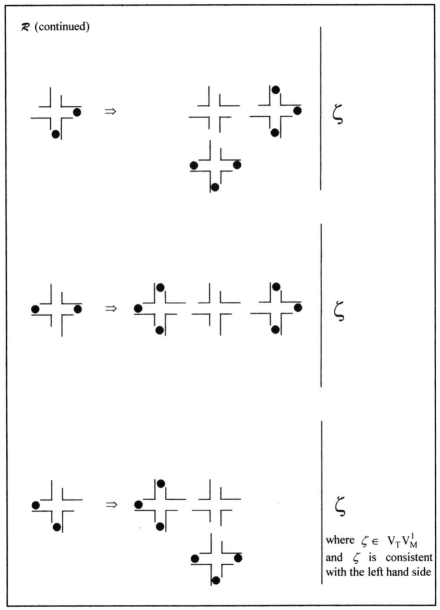

Figure A.7 Rules in $\mathcal{R}$ for Holliday shape grammar

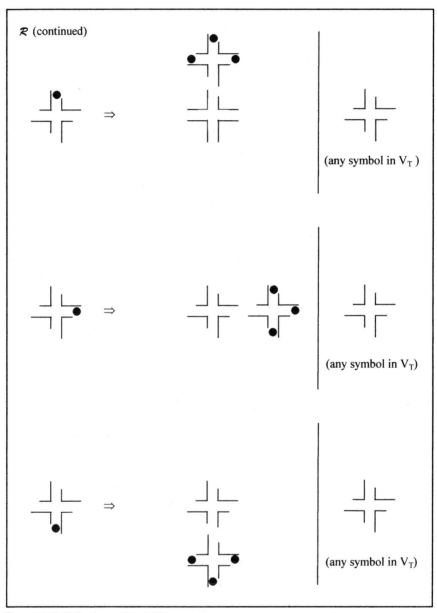

Figure A.8 Rules in $\mathcal{R}$ for Holliday shape grammar

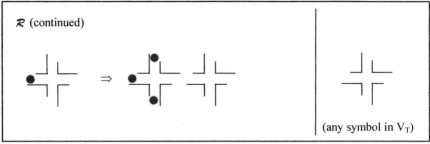

Figure A.9 Rules in $\mathcal{R}$ for Holliday shape grammar

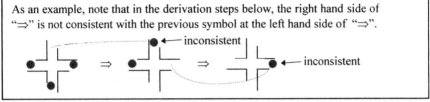

Figure A.10 Example of right hand sides being inconsistent with the previous symbol

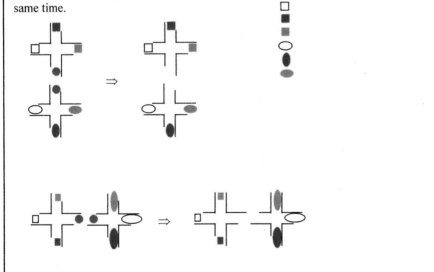

Figure A.11 General disambiguating rules in $\mathcal{R}$ for Holliday shape grammar

Examples of some (not all) disambiguating rules.

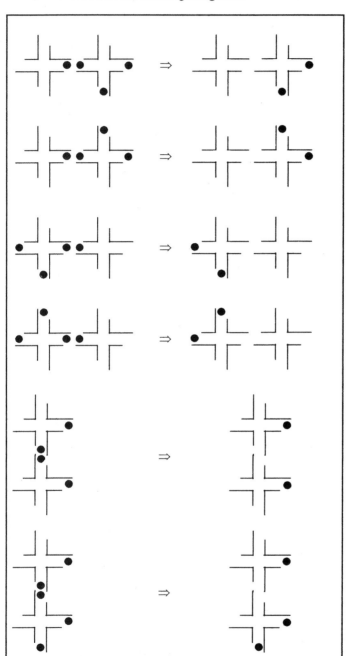

Figure A.12 Some specific disambiguating rules in $\mathcal{R}$ for Holliday shape grammar

A few derivations:
#1

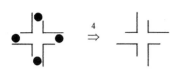

#2

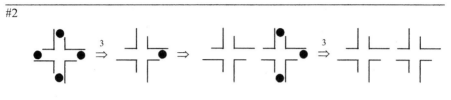

#3

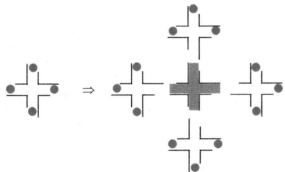

Each terminal "Holliday" structure is shown in a different shade to emphasize that the particular ordering and constituent bases in these structures is distinct as each may be probabalistically selected from a class of structures in V_T. This shape grammar may or may not be viewed probabalistically.

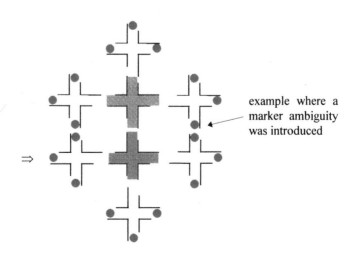

example where a marker ambiguity was introduced

disambiguate

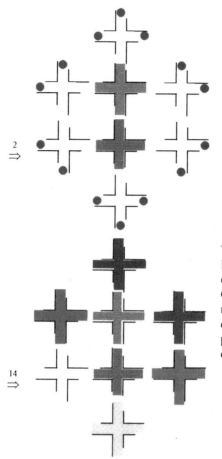

The possibly different Holliday structures in V_T could include iso-C/iso-G pairs, abasic pairs, mismatched pairs, self complementing bases, permutations of base orderings, etc.

If there is more than one element in V_T, then any convenient probability distribution may be chosen, or a deterministic shape grammar may be constructed. An example of a parallel shape grammar derivation step follows.

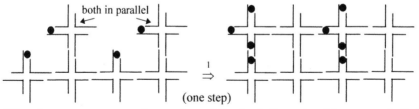

(one step)

Other shape grammars may be constructed for the 2-dimensional language of all planar networks of ligated Holliday structures. It is easy to construct a far simpler shape grammar for Holliday structures: a shape grammar that is very similar to the shape grammar for replication forks (which follows). Shape grammars are appropriate in this application, as the exact base structure (bases used and their sequences) may be exactly specified for all the finite number of species in V_T. Shape grammars could be a very useful tool.

The possibilities abound, such as ligating 2-dimensional branching structures typically observed as replication forks (provided they can be made to be sufficiently stable).

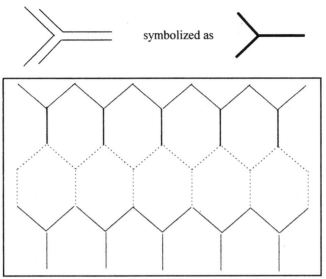

symbolized as

Figure A.13 Beginning the branching replication fork shape grammar

$$SG_{REP\ Y} = \left(V_M,\ V_T,\ \mathcal{R},\ I\right) \text{ where } V_M = \left\{ \bullet, \blacksquare, \square \right\}, V_T = \left\{ Y \right\}, I = \left\{ Y \right\}$$

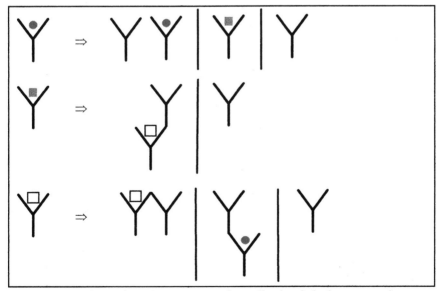

Figure A.14 Rules $\mathcal{R}$ for a shape grammar for replication fork planar networks

A sample derivation:

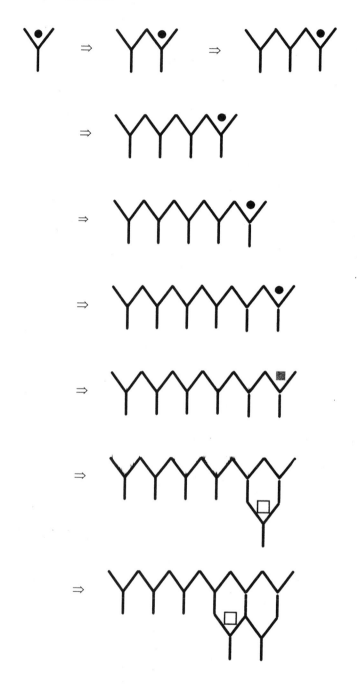

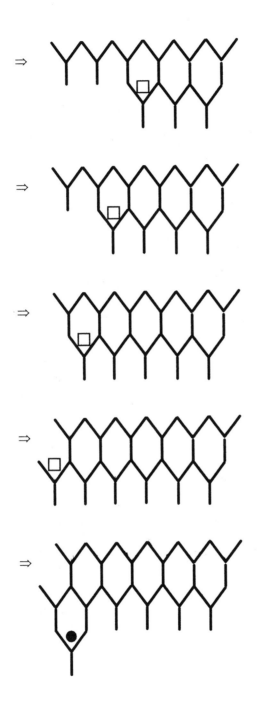

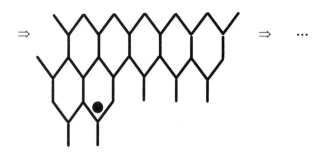

Recall that a shape grammar for 2-dimensional planar Holliday networks (which is much simpler than the one given can be constructed). Such a shape graamar is quite similar to the given replication fork shape grammar above.

Just as with Holliday structures, different orderings of bases, including artificial bases, mismatches, abasic pairs, etc. allow for a multiplicity of such replication fork structures, and this can be made to be probabilistic. It must be emphasized that for both the case of Holliday structures as well as replication forks, empirical evidence in the laboratory of the associated planar structures is lacking. Both these cases are idealizations, intended only to show how shape grammars might be useful as a linguistic tool.

Another possibility is G-quartets viewed as 3-dimensional shape grammars (G-quartets exist in extended forms at least in vitro, and are associated with telomeres, thus stability does not seem to be a problem).

Another possibility are shape grammars that will generate complicated RNA shapes including hairpins, clover-leafs, bulges, etc. Such 2-dimensional shape grammars could use associated Nussinov charts to place bases at corresponding sites. Intersecting chords that connect Nussinov "flowers" could (in 3-dimensions) underlie the links and knots found in tertiary RNA structures.

The last possibility to be mentioned is the application of shape grammars to biological materials based upon threose or hexose based analogues of DNA: the TNA discussed at the end of Chapter 3. However, not only would these structures have to be sufficiently stable, but in addition, ligases would then have to exist (either natural or artificial) that function with threose or hexose TNA molecules.

As a closing note, graph grammars and web grammars are just beginning to be applied to DNA, RNA, and proteins. These applications in their infancy and will not be discussed.

Technical References

1. Abrahams, J., van den Berg, M., van Batenburg, E., and Pleij, C.
    Prediction of RNA secondary structure, including pseudoknotting, by computer simulation
    Nucleic Acids Research, April 17, 1990, 18, 10, 3035-3044

2. Adleman, L. M.
    Molecular computation of solutions to combinatorial problems
    Science, 1994, 266, 1021-1024

3. Aho, A.
    Indexed grammars – An extension of context-free grammars
    Journal of the Association for Computing Machinery, 1968, 15, 4, 647-671

4   Alberti, P., Arimondo, P., Mergny, J., Garestier, T., Hélène, C., and Sun, J.
    A directional nucleation-zipping mechanism for triple helix formation
    Nucleic Acids Research, 2002, 30, 24, 5407-5415

5. Allison, L., Wallace, C., and Yee, C.
    Finite-state models in the alignment of macromolecules
    Journal of Molecular Evolution, 1992, 35, 77-89

6. Atkins, J., and Gesteland, R.
    The 22nd amino acid
    Science, May 24, 2002, 296, 1409-1410

7. Barnett, M.
    A molecular model for string transformations
    Association for Computing Machinery, SIGBIO Newsletter, 1985, 7, 4, 53-58

8. Basolo, F., and Pearson, R.
    Mechanisms of Inorganic Reactions
    Wiley, 1958

9. Benner, S., and Presnell, S.
    The design of synthetic genes,
    Nucleic Acids Research, 1988, 16, 5, 1693-1702

10. Benner, S., Switzer, C., and Moroney, E.
    Enzymatic incorporation of a new base pair into DNA and RNA
    Journal of the American Chemical Society, 1989, 111, 8322-8323

11. Benner, S., Piccirilli, J., Krauch, T., and Moroney, S.
    Enzymatic incorporation of a new base pair into DNA and RNA extends the genetic alphabet
    Nature, January 4, 1990, 343, 33-37

12. Benner, S., Switzer, C., and Moroney, E.
    Enzymatic recognition of the base pair between isocytidine and isoguanosine
    Biochemistry, 1993, 32, 10489-10496

13. Benner, S.
    Expanding the genetic lexicon: Incorporating non-standard amino acids into proteins
        by ribosome-based synthesis
    Trends in Biotechnology, May 1994, 12, 5, 158-163

14. Benner, S., Held, H., Hottiger, M., and Hübscher, U.
    Differential discrimination of DNA polymerases for variants of the non-standard
        nucleobase pair between xanthosine and 2, 4-diaminopyrimidine, two components
        of an expanded genetic alphabet
    Nucleic Acids Research, 1996, 24, 7, 1308-1313

15. Benner, S., Ciglic, M., Haugg, M., Jermann, T., Opitz, J., Raillard, S.,
    Stackhouse, J., Trabesinger-Rüf,N., Trautwein-Fritz,K., Weinhold, E.,
    and Zankel, T.
    Developing new synthetic catalysts. How nature does it
    Acta Chemica Scandinavica, 1996, 50, 243-248

16. Benner, S., Devine, K., Matveeva, L., and Powell, D.
    The missing organic molecules on Mars
    Proceedings of the National Academy of Sciences, March 14, 2000, 97, 6,
        2425-2430

17. Benner, S., Gaucher, E., and Miyamoto, M.
    Function-structure analysis of proteins using covariant-based evolutionary
        approaches: Elongation factors
    Proceedings of the National Academy of Sciences, January 16, 2000, 98, 2, 548-552

18. Benner, S., Chamberlin, S., Govindarajan, S., Knecht, L., and Liberles, D.
    Functional inferences from reconstructed evolutionary biology involving rectified
        databases-an evolutionary grounded approach to functional genomics
    Research Microbiology, March, 2000, 151, 2, 97-106

19. Benner, S., Hansen, P., Liberles, D., Peltier, M., and Raley, L.
    Evolutionary history of the uterine serpines
    Journal of Experimental Zoology, August 15, 2000, 288, 2, 165-174

20. Benner, S., and Gaucher, E.
    Evolution, language and analogy in functional genomics
    Trends in genetics, July, 2001, 17, 7, 414-418

21. Bentolila, S.
    A grammar describing "biological binding operators" to model gene regulation
    Biochimie, 1996, 78, 335-350

22. Blattner, F., and Schroeder, J.
    Formal description of a DNA oriented computer language
    Nucleic Acids Research, 1982, 10, 1, 69-85

23. Boyd, J.
    The algebra of group kinship
    Journal of Mathematical Psychology, 1969, 6, 139-167

24. Brendel, V., and Busse, H.
    Genome structure described by formal languages
    Nucleic Acids Research, 1984, 12, 5, 2561-2568

25. Brendel, V., Beckmann, J., and Trifonov, E.
    Linguistics of nucleotidesSequences: Morphology and comparison of vocabularies
    Journal of Biomolecular Structure & Dynamics, 1986, 4, 1, 011-021

26. Bukhari, A., Shapiro, J., and Adhya, S.
    DNA: Insertion Elements, Plasmids, and Episomes
    Cold Spring Harbor Laboratory, 1977

27. Burks, C., and Farmer, D.
    Towards modeling DNA sequences as automata
    Physica, 1984, 10, D, 157-167.

28. Chamberlin, A., Bain, J., Glabe, C., and Dix, T.
    Biosynthetic site-specific incorporation of a non-natural amino acid into a
        polypeptide
    Journal American Chemical Society, 1989, 111, 8013-8014

29. Chamberlin, A., Bain, J., Wacker, D., and Kuo, E.
    Site-specific incorporation of non-natural residues into polypeptides: Effect of
        residue structure on suppression and translation efficiencies
    Tetrahedron, 47, April 8, 1991, 14/15, 2389-2400

30. Chastain, M., and Tinoco, I.
    Structural elements in RNA
    Progress in Nucleic Acid Research and Molecular Biology, 1991, 41, 131-177

31. Chomsky, N., Hauser, M., and Fitch, W.
    The faculty of language: What is it, who has it, and how did it evolve?
    SCIENCE, November 11, 2002, 298, 1569-1579.

32. Collado-Vides, J.
    A transformational-grammar approach to the study of the regulation of gene
        expression
    Journal Theoretical Biology, 1989, 136, 403-425

33. Collado-Vides, J.
Towards a grammatical paradigm for the study of the regulation of gene expression.
In Theoretical Biology, Epigenetic and Evolutionary Order from Complex Systems,
B. Goodwin and P. Saunders, eds, Edinburgh University Press, 1989, 211-224

34. Collado-Vides, J.
The search for a grammatical theory of gene regulation is formally justified by
showing the inadequacy of context-free grammars
Computer Applications in the Biosciences (CABIOS), 1991, 7, 3, 321-326

35. Collado-Vides, J.
A syntactic representation of units of genetic information–A syntax of units of
genetic information
Journal Theoretical Biology, 1991, 148, 1, 401-429

36. Collado-Vides, J.
Grammatical model of the regulation of gene expression,
Proceedings of the National Academy of Sciences USA, 1992, 89, 9405-9409.

37. Collado-Vides, J.
A linguistic representation of the regulation of transcription initiation. I. An ordered
array of complex symbols with distinctive features
BioSystems, 1993, 29, 87-104.

38. Collado-Vides, J.
A linguistic representation of the regulation of transcription initiation. II. Distinctive
features of sigma 70 promoters and their regulatory binding sites
BioSystems, 1993, 29, 105-128.

39. Collado-Vides, J.
Towards a unified grammatical model of $\sigma 70$ and $\sigma 54$ bacterial promotors
Biochimie, 1996, 78, 351-363

40. Collado-Vides, J., Magasanik, B., and Smith, T.
Integrative representations of the regulation of gene expression, Chapter 9 of A
Linguistic approach to the Study of the Regulation of Gene Expression
MIT Press, 1996, 179-203.

41. Collado-Vides, J., Rosenblueth, D., Thieffry, D., Huerta, A., and Salgado, H.
Syntactic recognition of regulatory regions in Escherichia coli
Computer Applications in the Biosciences (CABIOS), 1996, 12, 5, 415-422

42. Courcelle, J., Chow, K.H., Courcelle, C., and Donaldson, J.
DNA damage-induced replication fork regression and processing in
Escherichia coli
Science, February 14, 2003, 299, 1064 -1067

43. Culik, K., and Harju, T.
    Dominoes and the regularity of DNA splicing languages
    Lecture Notes in Computer Science 352, Springer-Verlag, 1989, 222-233

44. Culik, K., and Harju, T.
    Splicing semigroups of dominoes and DNA
    Discrete Applied Mathematics, 1991, 31, 261-277

45. Dayhoff, M.
    Computer analysis of protein evolution
    Scientific American, 1969, 221, 1, 86-95

46. Dervan, P., Povsic, T., and Strobel, S.
    Sequence-specific double-strand alkylation and cleavage of DNA mediated by
        triple-helix formation
    Journal of the American Chemical Society, 1992, 114, 5934-5941

47. Devlin, T.
    Textbook of Biochemistry with Clinical Correlations
    J. Wiley, 2001.

48. Ellington, A., and Kumar, P.
    Artificial evolution and natural ribozymes
    FASEB Journal, September, 1995, 9, 12, 1183-1105

49. Ellington, A., McGrew, S., and Rozen, D.
    Does DNA compute?
    Current Biology, 1996, 6, 3, 254-257

50. Ellington, A., Hesselberth, J., and Robertson, M.
    Optimization and optimality of a short ribozyme ligase that joins non-Watson-Crick
        base pairings
    RNA, April, 2001, 7, 4, 513-523

51. Eschenmoser, A., Guntha, S., Krishnamurthy, R., Schöning, K., and Scholz, P.
    Chemical etiology of nucleic acid structure: The α-threofuranosyl-($3^1 \to 2^1$)
        oligonucleotide system
    Science, November 17, 200, 290, 1347-1351

52. Fararo, T.
    Mathematical Sociology: An Introduction to Fundamentals
    John Wiley, 1973, 191-200

53. Freund, R.
    Splicing Systems on Graphs
    Proceedings of INBS-IEEE Conference, 1995, 189-194

54. Gann, D., Ostrander, L., and Schoeffler, J.
    A finite state model for the control of adrenal cortical steroid secretion
    Systems Symp. (Third), Case Inst. of Tech. Syst. Theory and Biology,
    1968, 185-200

55. Garnieri, F., Fliss, M., and Bancroft, C.
    Making DNA add
    Science, 1996, 273, 220-223.

56. Garzon, M., Gao, Y., Rose, J., Murphy, R., Deaton, R., Franceschetti, D., and
    Stevens, S.
    In vitro implementation of finite-state machines
    Proceedings of the Second Annual Genetic Programming Conference, 1997, 56-71

57. Garzon, M., Gao, Y., Rose, J., Murphy, R., Deaton, R., Franceschetti, D., and
    Stevens, S.
    DNA implementation of finite-state machines
    Proceedings of the Second Annual Genetic Programming Conference, 1997,
        479-487

58. Garzon, M., Gao, Y., Rose, J., Murphy, R., Deaton, R., Franceschetti, D., and
    Stevens, S.
    In vitro implementation of finite-state machines
    Second International Workshop on Implementing Automata, 1997,
    Lecture Notes in Computer Science 1436, Springer-Verlag, 1997, 56-71

59. Gatterdam, R., and Denninghoff, K.
    On the undecidability of splicing systems
    International Journal of Computer Mathematics, 1989, 27, 133-145

60. Gatterdam, R.
    Splicing systems and regularity
    International Journal of Computer Mathematics, April 1, 1989, 31, 63-67

61. Gatterdam, R.
    DNA and twist free splicing systems
    Proceedings of the Int. Conference Words, Languages and Combinatorics,
        1992, 170-178

62. Gatterdam, R.
    Algorithms for splicing systems
    SIAM Journal of Computing, 192, 21, 3, 507-520

63. Goheen, H., and Stahl, W.
    Molecular algorithms
    Journal of Theoretical Biology, 1963, 5, 266-287

64. Head, T.
    Formal language theory and DNA: An analysis of the generative capacity of
        specific recombinant behaviors
    Bulletin of Mathematical Biology, 1987, 49, 6, 737-759

65. Head, T.
    Splicing schemes and DNA
    Lindenmayer Systems
    Springer-Verlag, 1992, 371-383

66. Head, T., Paŭn , G., and Pixton, D.
    Language theory and molecular genetics: Generative mechanisms suggested by
        DNA recombination
    Handbook of Formal Languages Linear Modeling: Background and Application
    Rozenberg, G., and Salomaa, A., eds., Springer-Verlag, January 2001, 2, 295-360

67. Head, T., Hagiya, M., and Clote, P.
    Computing with biomolecules
    Pacific Symposium on Biocomputing, 1998, 521-522

68. Holcombe, W.
    Algebraic Automata Theory
    Cambridge University Press, August 19, 1982

69. Huntsberger, T. and Park, K.
    Inference of context free grammars for syntactic analysis of DNA sequences
    Artificial Intelligence and Molecular Biology, 1990 Spring Symposium Series,
        110-114

70. Jiménez-Montaño, M., and Ebeling, W.
    On grammars, complexity, and information measures of biological macromolecules
    Mathematical Biosciences, 1980, 52, 53-71

71. Jiménez-Montaño, M.
    On the syntactic structure of protein sequences and the concept of grammar
        complexity
    Bulletin of Mathematical Biology, 1984, 641-659

72. Jiménez-Montaño, M.
    Formal languages and theoretical molecular biology
    Theoretical Biology Epigenetic and Evolutionary Order from Complex Systems
    Saunders, P. and Goodwin, B., eds., edinburgh University, April, 1990, 199-210

73. Joshi, A., Vijay-Shanker, K., and Weir, D.
    The Convergence of mildly context-sensitive grammar formalisms.
        In Foundational Issues in Natural Language Processing
    P. Sells, S. Shieber, T. Wasow, eds., MIT Press, 31-81

74. Kaminuma, T., and Matsumoto, G.
    Biocomputers: The Next Generation from Japan,
    Chapman and Hall, 1991, 42-44

75. Kim, S.
    Computational modeling for genetic splicing systems
    SIAM Journal of Computing, 1997, 26, 5, 1284-1309

76. Kobayashi, S., Ishida, N., and Yokomori, T.
    Learning local languages and its application to protein α-chain identification
    Proceedings of the Twenty-Seventh Annual Hawaii Int. Conference on System
        Science V, 1994, 113-122

77. Kobayashi, S., and Yokomori, T.
    Learning local languages and their application to DNA sequence analysis
    IEEE Transactions on Pattern Analysis and Machine Intelligence, 1998, 20, 10,
        1067-1079

78. Kobayashi, S., and Yokomori, T.
    DNA evolutionary linguistics and RNA structure modeling: A computational
        approach
    First Int. Symposium on Intelligence in Neural and Biological Systems, 1995, 38-45

79. Kobayashi, S., and Feretti, C.
    DNA splicing systems and Post systems
    Pacific Symposium on Biocomputing, 1996, 288-299

80. Kool, E., Liu, D., and Moran, S.
    Bi-stranded, multisite replication of a base pair between diflourotoluene and
        adenine: Confirmation by 'inverse' sequencing
    Chemical Biology, December, 1997, 4, 12, 919-926

81. Kool, E., and Morales, J.
    Efficient replication between non-hydrogen-bonded nucleoside shape analogs
    Nature Structural Biology, November 5, 1998, 5, 11, 950-953

82. Kool, E., Guckian, K., and Krugh, T.
    Solution structure of a DNA duplex containing a replicable diflourotoluene-adenine
        pair
    Nature Structural Biology, November 5, 1998, 5, 11, 954-959

83. Kool, E., Barsky, D., and Colvin, M.
    Interaction and solvation energies of nonpolar DNA base analogues and their role
        in polymerase insertion fidelity,
    Journal of Biomolecular Structure & Dynamics, 1999, 16, 6, 1119-1134

84. Kool, E., and Matray, T.
A specific partner for abasic damage in DNA
Nature, June 17, 1999, 399, 704-708

85. Kool, E., Pedroso, E., and Frieden, M.
Tightening the belt on polymerases: Evaluating the physical constraints on enzyme substrate size
Angew Chemical International Edition, 1999, 38, 24, 3654-3657

86. Kool, E., Guckian, K., and Morales, J.
Mimicking the structure and function of DNA: Insights into DNA stability and replication
Angew Chemical International Edition, 2000, 39, 6, 990-1009

87. Kool, E.
Hydrogen bonding, base stacking, and steric effects in DNA replication
Annual Revue of Biophysics and Biomolecular Structure, 2001, 30, 1-22

88. Kool, E., and Morales, J.
Importance of terminal base pair hydrogen-bonding in 3^1-end proofreading by Klenow fragment of DNA polymerase I
Biochemistry, March 14, 2000, 39, 10, 2626-2632

89. Kool, E.
Replacing the nucleobases in DNA with designer molecules
Accounts of Chemical Research, 2002, 35, 11, 936-943

90. Kornberg, A.
DNA Replication,
W.H. Freeman and Company, 1980

91. Krohn, K., Langer, R., and Rhodes, J.
A theory of finite physics with an application to the analysis of metabolic systems
Proceedings of the 1966 Bionics Symposium,
Cybernetic problems in bionics,
Gordon and breach Science, May 3-5, 1966, 633-648

92. Krohn, K., Langer, R., and Rhodes, J.
Transformations, Semigroups, and Metabolism,
Systems Theory & Biology, 3rd Systems Symposiums,
Proceedings–System Symposium (Case Institute of Terchnology)
Wiley, 1966, 130-140

93. Lathrop, R., Webster, T., and Smith, T.
Ariadne: Pattern-directed inference and hierarchical abstraction in protein structure
recognition
Communications of the ACM, 1987, 30, 11, 909-921

94. Lewin, B.
Gene Expression-1, Bacterial Genomes
Wiley, 1974

95. Lewin, B.
Genes IV
Oxford University Press, 1990

96. Lida, Y., Kudo, M., and Shimbo, M.
Syntactic pattern analysis of 5'-splice site sequences of mRNA precursors in
higher eukaryote genes
Computer Applications in the Biosciences (CABIOS), 1987, 3, 4, 319-324

97. Lindenmayer, A., and Jürgensen, H.
Grammars of development: Discrete-state models for growth, differentiation, and
gene expression in modular organisms
Lindenmayer Systems
Springer-Verlag, 1992, 3-5

98. Longshaw, T.
Evolutionary learning of large grammars
Proceedings of the Second Annual Genetic Programming Conference, 1997,
445-453

99. Mac Dónaill, Dónall A.
A parity code interpretation of nucleotide alphabet composition
Chemical Communications, 2002, 2062-2063

100. Magarshak, Y., and Benham, C.
An algebraic representation of RNA secondary structures
Journal of Biomolecular Structure and Dynamics, 1992, 10, 3, 465-488

101. Magarshak, Y.
Quaternion representation of RNA sequences and tertiary structures
BioSystems, 1993, 30, 21-29.

102. Magarshak, Y., Kister, A., and Malinsky, J.
The theoretical analysis of the process of RNA molecule self-assembly
BioSystems, 1993, 30, 31-48

103. Marcus, S.
Linguistic structures and generative devices in molecular genetics
C.L.T.A, 1974, 11, fasc. 1, Bucharest, 77-104

104. Martindale, C.
The grammar of the tic in Gilles de La Tourette's syndrome
Language and Speech, 1976, 19, 3, 266-275

105. Mian, I., Sakakibara, Y., Brown, M., Hughey, R., Sjölander, K., Underwood, R.,
and Haussler, D.
Stochastic context-free grammars for tRNA modelling
Nucleic Acids Research, 1994, 22, 23, 5112-5120

106. Mian, I., and Noller, H.
RNA modeling using Gibbs sampling and stochastic context free grammars
Proceedings Second International Conference on Intelligent Systems for Molecular
    Biology, R. Altman, D. Brutlag, P. Karp, R. Lathrop, D. Searls, eds.,
    1994, 138-146

107. Mitrana, V.
Crossover systems. A generalization of splicing systems
Journal of Automata, Languages and Combinatorics 2, 1997, 3, 151-160

108. Mitrana, V., and Dassow, J.
Splicing grammar systems
Computers and Artificial Intelligence, 1996, 15, 2-3, 109-122

109. Mitrana, V., and Dassow, J.
Evolutionary grammars: A grammatical model for genome evolution
Bioinformatics, German Conference on Bioinformatics, 1996, 199-209

110. Mueller, U., Maier, G., Onori, A., Cellai, L., Heumann, H., and Heinemann, U.
Crystal structure of an 8-base-pair fragment containing the 3'-DNA-5' junction
    formed during initiation of minus-strand synthesis of HIV replication
Biochemistry, 1998, 37, 35, 12005-12011

111. Naumann, B., Plank, F., Hofbauer, G.
Language and Earth: Elective affinities between the emerging sciences of linguistics
    and geology
John Benjamins, 1992

112. Noller, C.
Chemistry of Organic Compounds
Saunders, 1960

113. Nussinov, R., Pieczenik, G., Griggs, J., and Kleitman, D.
     Algorithms for loop matchings
     SIAM Journal of Applied Mathematics, 1978, 35, 1, 68-82

114. Nussinov, R., and Jacobson, A.
     Fast algorithm for predicting the secondary structure of single-stranded RNA
     Proceedings of the National Academy of Sciences USA, 1980, 77, 11, 6309-6313

115. Nussinov, R., and Tinoco, I.
     Sequential folding of a messenger RNA molecule
     Journal of Molecular Biology, 1981, 151, 519-533.

116. Oliver, J.
     Computation with DNA-matrix multiplication
     Second Annual Meeting on DNA based computers, DIMACS, Princeton University,
        American Mathematics Society, 1996, 236-248.

117. Orgel, L.
     A simpler nucleic acid
     Science, November 17, 2000, 290, 1306-1307

118. Ptashne, M., and Gann, A.
     Activators and targets
     Nature, 1990, 346, 329-331

119. Pattee, H.
     On the origin of macromolecular sequences
     Biophysical Journal, 1961, 1, 683-710

120. Păun, Gh.
     On the power of the splicing operation
     International Journal of Computer Mathematics, 1995, 59, 27-35

121. Păun, Gh.
     The splicing operation on formal languages
     First Int. Symposium on Intelligence in Neural and Biological Systems,
        1995, 176-180

122. Păun, Gh.
     Splicing: A challenge for formal language theorists
     Bulletin EATCS 1995, 57, 183-195

123. Păun, Gh.
     On the power of splicing grammar systems
     Analele Univ. Bucuresti Mathematica, 1996, XLV, 1, 93-96

124. Păun, Gh.
On the splice operation
Discrete Applied Mathematics, 1996, 70, 57-79

125. Păun, Gh.
Universal DNA computing models based on the splicing operation
Discrete Math. and Theoretical Computer Science (DIMACS), 1996, 44, 59-75

126. Păun, Gh.
Regular extended H systems are computationally universal
Journal of Automata, Languages and Combinatorics, 1996, 1, 1, 27-36

127. Păun, Gh., and Salomaa, A.
DNA computing based on the splicing operation
Math. Japonica, 1996, 43, 3, 607-632

128. Păun, Gh., and Salomaa, A.
From DNA recombination to DNA computing via formal languages
Bioinformatics, German Conference on Bioinformatics, 1996, 210-220

129. Păun, Gh., Rozenberg, G., and Salomaa, A.
Computing by splicing
Theoretical Computer Science, 1996, 168, 321-336

130. Păun, Gh., Rozenberg, G., Salomaa, A.
Restricted use of the splicing operation
International Journal of Computer Mathematics, 1996, 60, 17-32

131. Păun, Gh., Salomaa, A., and Kari, Lila
The power of restricted splicing with rules from a regular language
Journal of Universal Computer Science, 1996, 2, 4, 224-240

132. Păun, Gh., Csuhaj-Varjú, E., Freund, R., and Kari, L.
DNA computing based on splicing: Universality results
Pacific Symposium on Biocomputing, 1996, 179-190

133. Păun, Gh., Freund, R., Rozenberg, G., and Salomaa, A.
Bidirectional sticker systems
Pacific Symposium on Biocomputing, 1998, 535-546

134. Păun, Gh., Rozenberg, G., Salomaa, A., and Mateescu, A.
Simple splicing systems
Discrete Applied Mathematics, 1998, 84, 145-163

135. Păun, Gh., and Dassow, J.
Regulated rewriting in formal language theory
Springer-Verlag, 1989

136. Pereira, F., and Warren, D.
Definite clause grammars for language analysis –A survey of the formalism and
a comparison with augmented transition networks
Artificial Intelligence, 1980, 13, 1, 2, 231-278

137. Pixton, D.
Regularity of splicing languages
Discrete Applied Mathematics, 1996, 69, 101-124

138. Pixton, D.
Linear and circular splicing systems
First Int. Symposium on Intelligence in Neural and Biological Systems,
1995, 181-188

139. Pleij, C., Rietveld, K., and Bosch, L.
A new principle of RNA folding based upon pseudoknotting
Nucleic Acids Research, 1985, 13, 5, 1717-1731

140. Pleij, C., and Bosch, L.
RNA pseudoknots: Structure, detection, and prediction
Methods in Enzymology, 1989, 180, 289-303

141. Priese, L., Margenstern, M., and Rogojine, Y.
Finite H-systems with 3 test tubes are not predictable
Pacific Symposium on Biocomputing, 1998, 547-558

142. RajBhandary, U., Köhrer, C., and Kowal, A.
Twenty-first aminoacyl-tRNA synthetase-suppressor tRNA pairs for possible use in
site-specific incorporation of amino acid analogues into proteins in eukaryotes
and in eubacteria
Proceedings of the National Academy of Sciences, Feb. 27, 2001, 98, 5, 2268-2273

143. Rodger, R., and Rosebrugh, R.
Computing a grammar for sequences of behavioural acts
Animal Behaviour, 21979, 7, 737-749

144. Rothemund, P., and Smith, W.
DNA based computers
Discrete Math. and Theoretical Computer Science, 1996, 27, 76-184

145. Salomaa, A.
Formal Languages
Academic Press, 1973

146. Salomaa, A., and Mihalache, V.
Language-theoretic aspects of string replication
International Journal of Computer Mathematics, 1998, 66, 163-177

147. Salomaa, A., and Mihalache, V.
Growth functions and length sets of replicating systems
Acta Cybernetica, 1996, 12, 3, 235-247

148. Savitch, W.
A formal model for context-free languages augmented with reduplication
Computational Linguistics, 1989, 15, 4, 250-261

149. Schimmel, P., Crécy-Lagard, V., Döring, V., Hendrickson, T., Marlière, P.,
Mootz, H., and Nangle, L.
Enlarging the amino acid set of Escherichia coli by infiltration of the valine coding
pathway
Science, April 20, 2001, 292, 501-504

150. Schultz, P., Romesberg, F., Berger, M., McMinn, D., Ogawa, A., and Wu, Y.
Universal bases for hybridizaton, replication and chain termination
Nucleic Acids Research, 2000, 28, 15, 2911-2914

151. Schultz, P., Romesberg, F., Tolman, W., Holland, P., and Meggers, E.
A novel copper-mediated DNA base pair
Journal of the American Chemical Society, 2000, 122, 10714-10715

152. Schultz, P., Brock, A., Herberich, B., and Wang, L.
Expanding the genetic code of Escherichia coli,
Science, April 20, 2001, 292, 498-500

153. Schultz, P., Meggers, E., Spraggon, G.
Structure of a copper-mediated base pair in DNA
Journal of the American Chemical Society, 2001, 123, 12364-12367

154. Searls, D.
Representing genetic information with formal grammars
Proceedings American Association for Artificial Intelligence, 1988, 2, 386-391

155. Searls, D.
Investigating the linguistics of DNA with definite clause grammars
Logic Programming, Proceedings of the North American Conference, 1989,
189-208

156. Searls, D., Cheever, E., Karunaratne, W., and Overton, G.
Using signal processing techniques for DNA sequence comparison
Proceedings of the Fifteenth Annual Northeast Bioeng. Conference, 1989, 173-174

157. Searls, D.
Towards a computational linguistics of DNA
Artificial Intelligence and Molecular Biology, 1990 Spring Symposium, 122-126

158. Searls, D.
The computational linguistics of biological sequences, in
Artificial Intelligence, Hunter , ed., Chapter 2, 1993, 47-120

159. Searls, D., and Noordewier, M.
Pattern-matching search of DNA sequences using logic grammars
Conference on Artificial Intelligence Applications, 1991, 3 - 9

160. Searls, D.
The linguistics of DNA
American Scientist, 1992, 80, 579-591

161. Searls, D., and Dong, S.
A syntactic pattern recognition system for DNA sequences
The Second Int. Conference on Bioinformatics, Supercomputing and Complex
    Genome Analysis, 1992, 89-101

162. Searls, D.
String variable grammar: A logic grammar formalism for the biological
    language of DNA
The Journal of Logic Programming, 1955, 73-102

163. Searls, D., and Dong, S.
Gene structure prediction by linguistic methods
Genomics, 1994, 23, 540-551

164. Searls, D.
Formal grammars for intermolecular structure
First Int. Symposium on Intelligence in Neural and Biological Systems, 1995, 30-37

165. Shanon, B.
The genetic code and human language
Synthese, 1978, 39, 401-415

166. Simon, M.
Automata Theory
World Scientific, 1999

167. Siromoney, R., Dare, V., and Subramanian, K.
Circular DNA and splicing systems
Proceedings of Parallel Image Analysis, Lecture Notes in Computer Scence 654,
    Springer-Verlag, 1992, 260-273

168. Siromoney, R., and Takada, Y.
On identifying DNA splicing systems from examples
Analogical and Inductive Inference, International Workshop Proceedings, 1992,
    305-319

169. Smith, R.
A Finite state algo. for finding restriction sites and other pattern matching
     applications
Computer Applications in the Biosciences (CABIOS), 1988, 4, 4, 459-465

170. Stryer, L.
Biochemistry, 4th edition, 1995.
Freeman and Company, 1995

171. Szathmáry, Eörs
Four letters in the genetic alphabet: a frozen evolutionary optimum?
Proceedings of the Royal Society of London, Series B, 1991, 245, 91-99

172. Szybalski, W.
Universal restriction endonucleases: designing novel cleavage specificities by
     combining adapter oligodeoxynucleotide and enzyme moieties
Gene, 1985, 40, 169-173

173. Szybalski, W., Sun, K., Hasan, N., and Podhajska, A.
Class-IIS restriction enzymes--a review
Gene, 1991, 100, 13-26

174. Tanaka, K., and Shionoya, M.
Synthesis of a novel nucleoside for alternative DNA base pairing through metal
     complexion
Journal of Organic Chemistry, 1999, 64, 5002-5003

175. Tanaka, K., Yamada, Y., and Shionoya, M.
Formation of silver (I)-mediated DNA duplex and triplex through an alternative
     base pair of pyridine nucleobases
Journal of the American Chemical Society, 2002, 124, 8802-8803

176. Tanaka, K., Tengeiji, A., Kato, T., Shiro, M., Toyama, N., and Shionoya, M.
Efficient incorporation of a copper hydroxypyridone base pair in DNA
Journal of the American Chemical Society, 2002, 124, 12494-12498

177. Tanaka, K., Tengeiji, A., Kato, T., Toyama, N., and Shionoya, M.
A discrete self-assembled metal array in artificial DNA
Science, February 21, 2003, 299, 1212-1213

178. Tirrell, D., Fournier, M., Kawai, M., Mason, T., and McGrath, K.
Chemical and biosynthetic approaches to the production of novel polypeptide materials
Biotechnological Progress, 1990, 6, 188-192

179. Tirrell, D., Harden, J., McGrath, K., Petka, W., and Wirtz, D.
Reversible hydrogels from self-assembling artificial proteins
Science, July 17, 1998, 281, 389-392

180. Trifonov, E., Hirshon, J., and Pietrokovski, S.
    Linguistic measure of taxonomic and functional relatedness of nucleotide
        sequences
    Journal of Biomolecular Structure & Dynamics, 1990, 7, 6, 1251-1268

181. Waterman, M.
    Introduction to Computational Biology, Maps, Sequences and Genomes
    Chapman & Hall, 1996

182. Watson, J. D.
    Molecular Biology of the Gene
    Benjamin, 1965

183. Weizman, H., and Tor, Y.
    2, 2´-Bipyridine ligandoside: A novel building block for modifying DNA with
        intra-duplex metal complexes
    Journal of the American Chemical Society, 2001, 123, 3375-3376

184. Westman, R.
    Quantitative Methods in the Study of Animal Behavior
    Academic Press, 1977, 167-169, 177-179

185. White, A., Handler, P., and Smith, E.
    Principles of Biochemistry
    McGraw-Hill, 1968

186. Wolffe, A.
    Architectural transcription factors
    Science, 1994, 264, 1100-1101.

187. Zuker, M.   Sankoff, D.
    RNA secondary structures and their prediction
    Bulletin of Mathematical Biology, 1984, 46, 4, 591-62, References: Medical History

188. Chew, L.
    Geographic and individual variation in the morphology and sequential organization
        of the song of the Savannah sparrow (Passerculus sandwichensis)
    Canadian Journal of Zoology, 1981, 59, 4, 702-713

189. Chew, L.
    Finite state grammars for dialects of the advertising song of the Savannah sparrow
        (Passerculus sandwichensis)
    Behaviourial Processes, 1983, 8, 91-96

190. Davis, M.
    Computability & Unsolvability
    McGraw-Hill, 1958

191. Joshi, A., Levy, L., and Takahashi, M.
Tree adjoining grammars
Journal of Computer and System Sciences, 1975, 10, 1

192. Pulleyblank, D.
Of Topo and Maxwell's Dream
Science, August, 1997, 277, 648-649

193. Hailman, J., and Ficken, M. S.
Combinatorial animal communication with computable syntax: chick-a-dee calling
    qualifies as 'language' by structural linguistics
Animal Behaviour, 1986, 34, 1899-1901.

194. Hailman, J., Ficken, M. S., and Ficken, R. W.
Constraints on the structure of combinatorial 'chick-a-dee' calls
Ethology, 1987, 75, 62-80

195. Searcy, W. A.
Song repertoire and mate choice in birds
American Zoolology, 1992, 32, 71-80

Additional Technical References

196. Arkin, A., and Ross, J.
Computational functions in biochemical reaction networks
Biophysical Journal, 1994, 67, 560-578

197. Bennett, C.
Logical reversibility of computation
IBM Journal of Research and Development, 1973, 17, 6, 471-554

198. Bennett, C.
The thermodynamics of computation - a review
International Journal of Theoretical Physics, 1982, 21, 12

199. Borodovsky, M., Mironov, A., and Pevzner, P.
Linguistics of nucleotide sequences II: Stationary words in genetic texts and the
    zonal structure of DNA
Journal of Biomolecular Structure & Dynamics, 1989, 6, 5, 1027-1038

200. Buchner, A., and Funke, J.
Finite-state automata: Dynamic task environments in problem-solving research
The Quarterly Journal of Experimental Psychology, 1993, 46A, 1, 83-118

201. Conrad, M.
On design principles for a molecular computer
Communications of the Association for Computing Machinery, 1985, 28, 5, 464-480

202. Conrad, M.
Molecular computing
Advances in Computers, 1990, 31, 236-324

203. Fararo, T., and Skvoretz, J.
Languages and grammars of action and interaction: A contribution to the formal
theory of action
Behavioral Science, 1980, 25, 1, 9-22

204. Gann, D., Ostrander, L., and Schoeffler, J.
Identification of boolean mathematical models
Systems Symp. (Third), Case Inst. of Tech. Syst. Theory and Biology, 1968,
201-221

205. Hjelmfelt, A., Ross, J., and Weinberger, E.
Chemical implementation of finite-state machines
Proceedings of the National Academy of Sciences USA, 1992, 89, 383-387

206. Holcombe, W.
Towards a formal description of intracellular biochemical organization
Computers and Mathematical Applications, 1990, 20, 107-115

207. Kydd, R., and Wright, J.
Mental phenomena as changes of state in a finite-state machine
Australian and New Zealand Journal of Psychiatry20, 1986, 158-165

208. Lomnitz-Adler, J.
Automaton models of seismic fracture: Constraints imposed by the magnitude-
frequency relation
Journal of Geophysical Research, 1993, 98, 17745-17756

209. McClosky, J., Bean, C., Ren, J., and Steacy, S.
Heterogeneity in a self-organized critical earthquake model
Geophysical Research Letters, 1996, 23, 4, 383-386

210. Hailman, J., Ficken, M., and Ficken, R.
The 'chick-a-dee' calls of *Parus atricapillus*: A recombinant system of animal
communication compared with written English
Semiotica, 1985, 56, 3/4, 191-224

211. Hauser, M., and Konishi, M.
The design of animal communication. In
The Dance Language of Honeybees: Recent Findings and Problems,
by A. Michelsen,
MIT Press, 1999, 111-131

212. Herman, L., Richards, D., and Wolz, J.
Comprehension of sentences by bottlenosed dolphins
Cognition, 1984, 16, 129-219

213.. Martinez, H.
An automaton analogue on unicellularity
Biosysyems, 1979, 11, 133-162

214. Nelson, R.
Behaviorism, finite automata, and stimulus response theory
Theory and Decision, 1975, 6, 249-267

215. Paton, R., Nwana, H., Shave, M., and Bench-Capon, T.
An examination of some metaphorical contexts for biologically motivated
computing
The British Journal for the Philosophy of Science, 1994, 45, 2, 505-525

216. Roitblat, H., Herman, L., and Nachtgall, P.
Language and communication: Comparitive perspectives
Lawrence Erlbaum, 1993, Chapter 20, 430-455

217. Ruef, M.
Computational social grammar: Metatheory and methodology for exploring
structurational dynamics
Current Perspectives in Social Theory, 1997, 17, 129-157

218. Ruef, M.
Prolegomenon to the relation between social theory and method
Journal of Mathematical Sociology, 1997, 23, 3, 303-332

219. Saurin, W., and Marlière, P.
Matching relational patterns in nucleic acid sequences
Computer Applications in the Biosciences (CABIOS), 1987, 3, 2, 115-120

220. Schaffner, K.
The Watson-Crick model and reductionism
Topics in the Philosophy of Biology,
D. Reidel, January 1976, 27, 101-127

221. Shettleworth, S.
Cognition, Evolution, and Behavior,
Oxford University Press, 1998, 532 - 538.

222. Staden, R.
A computer program to search for tRNA genes
Nucleic Acids Research, 1980, 8, 4, 817-825
Chapter Medical History References

References Related to the Plague

223. Cipolla, C.
Faith, Reason, and the Plague in Seventeenth-Century Tuscany
W. W. Norton, 1981

224. Crosby, A.
The Colombian Exchange: Biological and Cultural Consequences of 1492
Greenwood Press, 1973

225. Dols, M.
The Black Death in the Middle East
Princeton university Press, 1977

226. Dubos, R., and Dubos, J.
The White Plague: Tuberculosis, Man, and Society
Rutgers University Press, 1987

227. Gottfried, R.
The Black Death: Natural and Human Disaster in Medieval Europe
Free Press, a Division of Macmillan, 1985

228. Horrox, R.
The Black Death
Manchester University Press, 1994

229. Lovell, W., and Cook, N.
"Secret Judgments of God" Old World Disease in Colonial Spanish America
University of Oklahoma Press, 1992

230. McNeill, W.
Plagues and Peoples
Doubleday Anchor, 1976

231. Siraisi, N.
Medieval & Early Renaissance Medicine
University of Chicago Press, 1999

232. Zinsser, H.
Rats, Lice and History: A Chronicle of Pestilence and Plagues
Black Dog & Leventhal, 1963

233. Brinton, D., ed.
The Annals of the Cakchiquels
Philadelphia, 1885

Index

Collating sequence for this index, in order:

shortest name before longer name,
Greek before other European alphabets
script
space,
special symbols : / etc.,
numbers,
alphabetics (case insensitive)